I0063858

Bulletin 59

DEPARTMENT OF THE INTERIOR

BUREAU OF MINES

JOSEPH A. HOLMES, Director

INVESTIGATIONS OF

DETONATORS AND ELECTRIC DETONATORS

BY

CLARENCE HALL

AND

SPENCER P. HOWELL

WASHINGTON
GOVERNMENT PRINTING OFFICE
1913

First edition.   June, 1913.

CONTENTS.

ILLUSTRATIONS.

INVESTIGATIONS OF DETONATORS AND ELECTRIC DETONATORS.

By Clarence Hall and Spencer P. Howell.

INTRODUCTION.

Among the more important factors involved in the use of high explosives in blasting operations is the means employed to bring about the detonation of the charge. When flame is applied to high explosives many of them may burn if not confined; but all of them when burning under certain conditions of confinement may detonate. Detonation may also be effected by mechanical means, such as frictional impact caused by a blow or by rubbing between surfaces. By this means, however, the full effect of the explosive charge may not be developed, so that a partial detonation, often accompanied by the burning of the explosive, results.

When nitroglycerin was first used it was fired by the application of flame, but considerable difficulty was experienced in exploding it with certainty and in obtaining uniform results. In 1864 Alfred Noble, a Swedish engineer, discovered that nitroglycerin could be surely and completely detonated by exploding in contact with it a small quantity of an initiatory explosive. Mercury fulminate was the substance then found capable of producing the best results There are many other fulminates and other substances that will produce complete detonation of commercial "high" explosives, but detonators or electric detonators containing mercury fulminate as the characteristic ingredient are still almost exclusively used in this country.

The term "detonator" is used in the publications of the Bureau of Mines to designate what the miner calls a "blasting cap"—a copper capsule containing a small quantity of some detonating compound that is ignited by a fuse. The term "electric detonator" is applied to a blasting cap that is similar except for being ignited by means of a small wire which is heated to incandescence or fused by the passage of an electric current.

One of the conditions prescribed by the Bureau of Mines for a permissible explosive [a] is that it shall be fired by a detonator, or preferably an electric detonator, having a charge equivalent to that of the standard detonator used at the Pittsburgh testing station. A further

---

[a] Permissible explosives have a short, quick flame and are intended especially for use in coal mines containing inflammable gases or dusts. (See Miners' Circular 6, Bureau of Mines.)

requirement is that this charge shall consist by weight of 90 parts of mercury fulminate and 10 parts of potassium chlorate (or their equivalents).

At the request of a manufacturer of permissible explosives, an investigation was undertaken by the bureau to determine the relative strength of detonators and electric detonators having different compositions. The tests of electric detonators herein reported were conducted by H. F. Braddock, junior chemist; J. W. Koster, J. E. Tiffany, junior mining engineers; and A. S. Crossfield, junior explosives chemist, at the Pittsburgh testing station of the bureau. Similar tests of detonators were not conducted because it was believed that the results would not show sufficient variation to warrant such tests. It is hoped that the conclusions drawn from the tests made will be of service to those using explosives by enabling them to select the grade of detonator or electric detonator that will insure the most effective results. The conclusions are given in this bulletin, which is published by the Bureau of Mines as one of a series of publications dealing with the testing of explosives and the precautions that should be taken to increase safety and efficiency in the use of explosives in mining operations.

The results of the experiments described in this bulletin show that the average percentage of failures of explosives to detonate was increased more than 20 per cent when the lower grades of electric detonators were used instead of No. 6 electric detonators, and was increased more than 50 per cent when these lower grades were used instead of No. 8 electric detonators. It is noteworthy, however, that when sensitive explosives, such as 40 per cent strength ammonia dynamite (p. 33), were tested under conditions ideal for detonation, the same energy was developed irrespective of the electric detonator used. When tests were made with a less sensitive explosive, such as a 40 per cent strength ammonia dynamite containing nitrosubstitution compounds (p. 32), the energy developed increased with the grade of the electric detonator used. For example, the average efficiency of four different explosives was increased 10.4 per cent when a No. 6 electric detonator was used instead of a No. 4 electric detonator, and 14.9 per cent when a No. 8 electric detonator was used (see tabulation on p. 45). The results of the tests emphasize the importance of using explosives in a fresh condition, but as fresh explosives can not always be had in mining work, strong detonators should be used in order to offset any deterioration of explosives from age.

The results obtained substantiate the following conclusions: (1) That for any particular manufacturer's detonators or electric detonators the explosive efficiency increases with their grade, and (2) that the four No. 6 electric detonators, of different makes, tested have practically the same explosive efficiency as, and each is considered equivalent to, the Pittsburgh testing station standard No. 6 electric detonator for use with permissible explosives in coal mines when the No. 6 grade is prescribed.

PRELIMINARY CONSIDERATIONS.

Methods for determining the strength of detonators or electric detonators by mechanical effects may be classed as either direct or indirect. The direct method comprises those tests in which the mechanical effect of the detonators or electric detonators is determined. The indirect method comprises those tests in which the mechanical effect of the explosives with which the detonators or electric detonators are used is determined. The direct method offers the advantage of simplicity, and usually of cheapness, but may lead to grave inaccuracies unless checked by mechanical effects indirectly determined. The discussion under the heading, "Tests Previously Used to Determine the Strength of Detonators" illustrates this.

The indirect method of determining the mechanical effects of explosives, or the energy developed by them, approximates practical conditions and offers an accurate means for determining the relative efficiency of detonators and electric detonators in bringing about complete detonation of commercial "high" explosives.

As all direct methods of testing detonators are therefore dependent on the indirect method for verification, the first experiments undertaken were to determine the relative strength of electric detonators indirectly by comparing the energy developed by different commercial explosives when fired with different grades of electric detonators. Afterwards tests were made by determining the relative strength of electric detonators by direct means, and a test was devised that, although not entirely satisfactory, gave results that approximated more closely those established by the indirect tests.

Detonating explosives develop their energy in the most efficient way when fired with detonators or electric detonators that completely detonate or explode them. Obviously, if the detonation be incomplete, a part of the potential energy of the explosive will not be released, and the loss of energy will be proportional to the percentage of the charge that did not detonate. In blasting operations an incomplete detonation is not only a menace to safety, by reason of the possible explosion of the unexploded part of the charge and of the harmful gaseous products resulting from the blast, but in many cases it acts like an underloaded shot and performs little, if any, useful work.

If an explosive is in a fresh condition and is sensitive to detonation and no obstacles are present to hinder its detonation, then any detonator effective enough to cause its complete detonation will develop its full energy.

In practice, however, conditions ideal for detonation rarely and perhaps never exist, because the commercial explosives are somewhat insensitive to detonation or because they may have deteriorated by aging before use.   Furthermore, crimped paper ends of the cartridges, loose material in the drill hole, air spaces between the cartridges, or cartridges of too small diameter may hinder detonation.

An explosive is said to age when any physical or chemical change during storage affects its sensitiveness, its uniformity, or its stability. Such changes are usually caused by the temperature and the humidity of the air or by sunlight, and even gravity may have an important effect.   If the explosive is placed in the sunlight, it may become unstable.   If a cartridge of dynamite is subjected to a temperature above 90° F., gravity may cause the segregation of nitroglycerin in the lower end or side of the cartridge.   If nitroglycerin explosives are subjected to temperatures alternately above and below 52° F., the nitroglycerin tends to segregate in the cartridge.   These conditions affect the uniformity of the explosives.   If explosives, especially those containing ammonium nitrate or other hygroscopic salts, are subjected to a moist atmosphere they tend to absorb moisture.   If the temperature is less than 52° F., nitroglycerin explosives other than low-freezing ones may freeze, and low-freezing explosives will freeze at a temperature less than 35° F.   Recently, there have been placed on the market nitroglycerin explosives, styled nonfreezing explosives, that are declared by the manufacturers to remain unfrozen when the temperature falls as low as 0° F.   Both moisture and freezing affect the sensitiveness of explosives.

The results of preliminary tests indicated that it would be impossible to discriminate between commercial electric detonators by determining the energy developed by explosives used with them unless the explosives were insensitive to detonation or tested under conditions which would simulate their use under actual mining conditions; consequently the authors, in testing electric detonators indirectly, used explosives in an insensitive condition.   This was done by using those that were naturally insensitive, such as an explosive of class 1,[a] subclass b; by using explosives in cartridges having small diameters, such as the 20 per cent "straight" nitroglycerin dynamite in cartridges of ⅞-inch diameter; in an aged condition, such as the 35 per cent strength gelatin dynamite two years old; in a frozen condition, such as 40 per cent strength gelatin dynamite; and, in the case of ammonia dynamite, by the addition of water.

The apparatus used in the experimental study were the Mettegang recorder, small lead blocks, and Trauzl lead blocks.   The results of the tests differentiated the electric detonators in two ways.   In the first place the electric detonator either did or did not cause the detonation of the explosives.   In tabulating such results the number of detonations is expressed as the percentage of the number of trials. Only the tests of those explosives were considered in which at least one failure to detonate occurred and in which detonation occurred in

---

at least one trial. In the second place, those trials in which detonation occurred were used as a basis of comparing the relative explosive efficiency of the electric detonators. The results of only those tests in which each of the electric detonators of the series caused detonation were recorded. The results are expressed in percentages of explosive efficiency as compared with the Pittsburgh testing station standard No. 6 electric detonator. This electric detonator offered the advantage of being included in both groups tested—the Pittsburgh testing station standard and the four No. 6 electric detonators.

THEORY OF DETONATION.

A short discussion of the theory of detonation as presented by Berthelot[a] is necessary in order that a better interpretation of the experiments herein reported can be made. The theory is called the "explosive-wave" theory, and it has been generally accepted because all detonation phenomena can be best explained by it. In order to analyze the propagation of an explosive wave, the wave is considered as a recurring cycle of released and transformed energy with four phases, as follows: Mechanical to calorific, calorific to chemical (the phase in which the potential energy of the explosive material is released), chemical to calorific, and calorific to mechanical.

This cycle can best be readily understood by indicating how the explosive wave is propagated through a cylindrical file of a homogeneous explosive without the loss of enough energy to interrupt propagation.

1. *Transformation of mechanical energy to calorific energy.*—When an explosive detonates a part of the mechanical energy of a layer of the explosive is converted instantly into heat energy in the adjacent layer by reason of the impact of molecules. The efficiency of this conversion is low—certainly less than 50 per cent—as the movement of the molecules is radial and they are only partly confined by the layer of explosive in the file. The mechanical energy that is not converted into heat energy exerts pressure on the confining medium and thus becomes the vehicle through which work is accomplished. There is good reason for believing that the thickness of the layer of explosive that enters into the first phase of the cycle varies with the physical properties of the explosive material, principally with its elasticity and partly with the velocity of the molecules that are in molecular vibration. The less elastic the explosive material and the greater the velocity of the molecules the thinner the layer, and hence the more times the cycle will recur in a unit length of the explosive material.

2. *Transformation of calorific energy to chemical energy.*—Some of the calorific energy of the layer is used to overcome the chemical stability of the explosive material, which may vary widely, and thus release the potential energy of the layer; the rest of the calorific

---

[a] Berthelot, M., Explosives and their power, 1892, pp. 88–113.

energy is used to accelerate and reinforce the chemical action. The layer of explosive by this time is developing a tremendous kinetic energy as expressed in phase 3.

3. *Transformation of chemical energy to calorific energy.*—All commercial explosives develop heat on detonation. This phase is different from the others because each of those represents some kind of kinetic energy derived entirely from the preceding phase, and consequently no one of them can have more kinetic energy than the preceding phase is capable of transferring. The conversion in this phase is complete because all the potential energy released becomes kinetic energy, which is largely calorific energy.

4. *Transformation of calorific energy to mechanical energy.*—A simple statement of this phase is that the larger volume of gases then formed from the layer of explosives is in an extremely active state of molecular vibration and that these molecules are then manifesting their energy as mechanical energy. The efficiency of conversion of calorific energy to mechanical energy is high because the conversion is very rapid and radiation and conduction losses are correspondingly small.

DETONATION OF HIGH EXPLOSIVES.

All methods used to initiate the explosive wave, or to detonate high explosives, involve the application of heat. If heat be applied directly by means of a flame such as is produced by a fuse, squib, or electric igniter, or by a spark or an incandescent solid, and the explosive be of the first order, or directly explosive, such as mercury fulminate or iodide of nitrogen, then detonation is sure and effective. If, however, the explosive be of the second order, or indirectly explosive, such as dynamite, permissible explosives, trinitrotoluene, or guncotton, then detonation, especially complete detonation, does not usually occur; hence the direct application of heat is not a sure and effective means of producing detonation.

If heat, such as is produced by the physical resistance of the explosive to a blow or impact, be applied indirectly to high explosives, then any sufficient blow or impact will cause detonation; that is, it will initiate the four-phase energy cycle, or explosive wave.

Because the impact produced by detonators is extremely quick, and their mercury-fulminate composition has a high density and releases considerable kinetic energy, the force of the impact is instantly converted into heat which is applied to a thin layer of the explosive material, thereby overcoming the chemical stability of that layer and initiating the explosive wave. Experience and investigation has proved this means of producing the detonation of explosives, those not too insensitive, to be both sure and effective; hence one is not surprised to learn that detonators are universally used.

As the mercury-fulminate composition of detonators is an explosive of the first order it may be detonated by fire, and hence fuse may be used in connection with them. Fuse is made of a uniform outside diameter and detonators are made of a uniform inside diameter such that the fuse fits snugly into them. In using fuse, it is cut square across and inserted into the detonator until it gently touches the fulminate mixture and then the detonator is crimped on the fuse.

Similarly a detonator may be fired by means of a small platinum wire embedded into the priming composition and brought to incandescence or fused by the passage of an electric current. (See figs. 1 and 4.) The priming composition may be simply an easily inflamed material such as loose guncotton, a match composition, an explosive of the first order such as mercury fulminate, or a mercury-fulminate composition. The priming composition is placed in the detonator directly above and in contact with the main charge. The platinum bridge is attached at each end to an insulated wire; the two wires, called the legs, pass through the plug and the filling, and are connected by leading wires to the source of the electric current. When a detonator is fitted with means of firing by an electric current it is called an electric detonator. Electric detonators are particularly adapted to shot firing in fiery mines, or to the simultaneous firing of several charges. They are also adapted to any purpose for which detonators may be used, and as their use offers a greater assurance of safety they are growing in favor.

ELECTRIC DETONATORS TESTED.

The electric detonators tested were designated as the Pittsburgh testing station standard No. 3, No. 4, No. 5, No. 6, No. 7, and No. 8, the Western Coast No. 6, the special No. 6, and the foreign No. 6. For brevity the expression Pittsburgh testing station standard is abbreviated in this paper to P. T. S. S.

The P. T. S. S. No. 3 electric detonators were made at the testing station from No. 3 detonators. A cross-sectional view of one of these electric detonators is shown in figure 1. The priming charge consisted of 0.02 gram of dry, loose guncotton directly above and in contact with the compressed charge. The sulphur plug, the insulated-wire legs, and the platinum bridge were so placed that the bridge was embedded in the loose guncotton. Then the molten sulphur was poured over the plug until the cap was filled.

As detonators in this country are made of a uniform inside diameter of 0.220 inch and electric detonators of a uniform inside diameter of 0.260 inch, the P. T. S. S. No. 3 electric detonators are smaller in diameter than all the others except the special No. 6 electric detonators which were also assembled at the Pittsburgh testing station. It was impossible to procure No. 3 electric detonators in the open market, as their manufacture has recently been discontinued.

The priming charge used in the No. 3, the No. 5, and the No. 7 electric detonators consisted of loose guncotton; that in the No. 4, the No. 6, and the No. 8 electric detonators was commercially pure mercury fulminate.

The Western Coast No. 6 and the foreign No. 6 electric detonators were used as received. The special No. 6 electric detonator was made at the testing station in the same manner as the P. T. S. S. No. 3. The primer of the western coast No. 6 was loose guncotton; that of the foreign No. 6 was a mixture of picric acid and chlorate of potash. The foreign No. 6 was so called because the detonator was imported, but the priming charge, sulphur plug, and wires were assembled by a manufacturer in this country.

These electric detonators are representative of all the electric detonators commercially used in the United States.

The P. T. S. S. No. 4, No. 5, No. 6, No. 7, and No. 8 were used as received from the manufacturers. Because of the seemingly erratic results of tests with the P. T. S. S. No. 5 electric detonators, attention is called to the fact that they were from 3 to $3\frac{1}{2}$ years old when used, and that although the sulphur plug protected the fulminating composition somewhat, they were not in first-class condition.

EXPLOSIVES USED IN THE TESTS.

The explosives used in the tests are enumerated below; they included certain permissible explosives and different grades of commercial dynamites. Explosives designated as permissible by the bureau are grouped in four classes.[a] Class 1, ammonium-nitrate explosives, includes all explosives in which the characterisitc material is ammonium nitrate. The class is divided into two subclasses: Subclass a, including every ammonium-nitrate explosive that contains a sensitizer that is itself an explosive, and subclass b, including every ammonium-nitrate explosive that contains a sensitizer that is not in itself an explosive. Class 2, hydrated explosives, includes all explosives in which salts containing water of crystallization are the characteristic materials. Class 3, organic-nitrate explosives, includes all explosives in which the characteristic material is an organic nitrate other than nitroglycerin. Class 4, nitroglycerin explosives, includes all explosives in which the characteristic material is nitroglycerin.

The permissible explosives used in the tests were as follows: Sample 1, sample 2, and sample 3 of an explosive of class 1, subclass a; sample 1 and sample 2 of an explosive of class 1, subclass b; and an explosive of class 4.

The commercial grades of dynamites used were a 20 per cent "straight" nitroglycerin dynamite; a 40 per cent strength ammonia dynamite (containing nitrosubstitution compounds); a 40 per cent strength ammonia dynamite; a 35 per cent strength gelatin dynamite (2 years old); a 35 per cent strength gelatin dynamite (3 years old); and a 40 per cent strength gelatin dynamite.

---

a See Miners' Circular 6, Bureau of Mines, 1912, p. 16.

The results of physical examination of the above-mentioned explosives were as follows:

Results of physical examination of explosives used in tests.

Class and grade of explosives.	Diameter of cartridge.	Length of cartridge.	Average weight.	Cartridges redipped.	Apparent specific gravity of cartridge by sand.	Color.	Consistence.
	In.	*In.*	*Gms.*				
Class 1, subclass *a* (sample 1).	1¼	8	160	No..	1.01	Corn	Granular and fibrous; fine; soft; dry; slightly cohesive.
Class 1, subclass *a* (sample 2).	1¼	8	174	Yes.	1.09	.....do	Do.
Class 1, subclass *a* (sample 3).	1½	8	227	Yes.	.93	Mauve........	Powdered; very fine; soft; dry; not cohesive.
Class 1, subclass *b* (sample 1).	1¾	8	277	Yes.	.88	Corn	Powdered; very fine; very dry; very soft; not cohesive.
Class 1, subclass *b* (sample 2).	1¾	8	278	Yes.	.88	.....do	Do.
Class 4..................	1¼	8	166	No..	1.00	.....do	Granular and fibrous; soft; fine; dry; slightly cohesive.
20 per cent "straight" nitroglycerin dynamite.	⅞	8	103	No..	1.18	.....do	Do,
40 per cent strength ammonia dynamite (containing nitrosubstitution compounds).	1¼	7⅞	226	Yes.	1.34	.....do	Fibrous; very fine; dry; soft; slightly cohesive.
40 per cent strength ammonia dynamite.	1¼	8	241	Yes.	1.43	Drab..........	Granular; fine; dry; soft; slightly cohesive.
35 per cent strength gelatin dynamite (2 years old).	1¼	7⅞	265	No..	1.63	Corn	Gelatinous; fine; wet; soft; moderately cohesive.
35 per cent strength gelatin dynamite (3 years old).	1⅜	7¾	339	No..	1.66	.....do	Do.
40 per cent strength gelatin dynamite.	1¼	7¾	295	No..	1.60	Drab..........	Do.

Certain of the different explosives used in the tests were analyzed, with results as follows:

Results of analyses of certain explosives used in tests.

Constituent.	Kind of explosives.					
	20 per cent "straight" nitroglycerin dynamite.[a]	40 per cent strength ammonia dynamite (containing nitrosubstitution compounds).[b]	40 per cent strength ammonia dynamite.[a]	35 per cent strength gelatin dynamite (3 years old).[b]	35 per cent strength gelatin dynamite (3 years old).[b]	40 per cent strength gelatin dynamite.[a]
Moisture............................	1.20	1.93	0.88	1.89	5.86	1.47
Nitroglycerin........................	19.54	16.28	21.60	29.03	28.10	30.70
Nitrololuene........................		4.97				
Nitrocellulose......................				.88	1.17	.88
Sodium nitrate.....................	c 62.09	47.14	46.04	48.62	52.20	54.27
Ammonium nitrate.................		18.78	18.86			
Wood pulp..........................	15.22		5.45		5.55	8.58
Wood pulp and crude fiber.........		2.84		2.15		
Calcium carbonate.................	1.95		1.44	1.13		
Zinc oxide..........................		.62	.88		1.07	1.02
Sulphur............................		2.84	4.85	4.83	4.58	3.08
Starch..............................		3.79		11.47		
Vaseline............................		.81				
Paraffin............................					1.24	
Rosin...............................					.23	
Total....................	100.00	100.00	100.00	100.00	100.00	100.00

[a] Analyst, W. C. Cope. [b] Analyst, A. L. Hyde. [c] Contains 1.04 per cent sodium chloride.

TESTS PREVIOUSLY USED TO DETERMINE STRENGTH OF DETONATORS AND ELECTRIC DETONATORS.

Six principal tests have been used previously to determine the strength of detonators or electric detonators. They are as follows:

1. *Weight of charge.*—Ever since it was observed that certain explosives would not always detonate with a certain weight of charge of mercury fulminate or mercury-fulminate composition and that these same explosives would always detonate if the weight of charge in the detonator was increased, it has been customary to vary the charge in the detonators and to consider the weight of the charge to be an indication of the strength of the detonator. There are several grades of detonators, and they are designated by the charge of fulminate composition contained in them.

Bigg-Wither [a] arranged the following table, which was published in 1900:

Weight of charge in different grades of detonators.

Grade No.	Charge per detonator.	
	Grams.	Grains.
1	0.30	4.6
2	.40	6.2
3	.54	8.3
4	.65	10.0
5	.80	12.3
6	1.00	15.4
6½	1.25	19.2
7	1.50	23.1
8	2.00	30.9

It is to be noted that in 1900 there was no great variation in the composition of detonators. There is no indication that the relation between the effectiveness of the detonator and the weight of the charge was other than directly as the first-power function.

2. *Deformation or penetration of lead or iron plates.*[b]—Guttman[c] states: "One of the oldest and most frequently used tests for measuring the power of caps (used only with ordnance) consisted of exploding them on a lead or iron plate resting on a hollow iron ring and estimating their strength from the deformation or the penetration of the block. For larger detonators of between one-half gram and 1 gram charge as used for borehole shots, the plate would have to be of greater thickness."

3. *Radial lines on lead plates.*—Bigg-Wither, in the article mentioned above, describes in considerable detail tests made with different detonators. He used lead plates 3 mm. thick for detonators Nos. 1 to 3 and lead plates 5 mm. thick for detonators Nos. 4 to 8.

---

[a] Bigg-Wither, H., Notes on detonators: Trans. Inst. Min. Eng., vol. 21, 1900, p. 442.

[b] Munroe, C. E., Lecture on chemistry and explosives, 1888, pp. 22–23.

[c] Guttman, Oscar, Manufacture of explosives, vol. 2, 1895, p. 369.

The lead plates were supported on the edges, and the detonators were placed vertically on the centers of the plates. He further states that after the tests the plates may be taken as direct pictorial records of the efficiency of the detonators but that they do not record the report of the explosion, the recording of which is essential; that the detonating effect is not shown so much by the punctures as by the fine radiating marks upon the surface of the plates; that the fine markings show that the force of the explosion smashes the copper tubing to powder, some of which often adheres to the sides of the plates, and that when there are fine radiating lines around the center there are heavier markings outside. The difference in effect is probably due to the upper part of the fulminate not being completely detonated. The results of tests show that detonators may absorb moisture when stored and emphasize the importance of using a detonator of higher power than would be otherwise actually requisite.

It appears, then, that this test is one that might readily be used to distinguish between good and poor or defective detonators regardless of the charge that they contain, and for this purpose the test appears to have considerable merit. However, as an indication of the relative effectiveness of detonators of different grades, that is, containing different weights of charge, it appears to have little value.

4. *Photographs of flashes from electric detonators.*—De Grave[a] conceived the idea that the flash or flame of a detonator might vary with the grade of the detonator, and such was the result of tests made by him. He also showed that there was little, if any, difference whether the electric detonator was of high or low tension. The following table gives the results for low-tension detonators:

Results of photographs of flashes of low-tension detonators.

Grade No.—	Dimension of flash.
	Inches.
3	1.0 by 0.22
6	1.6 by .22
7	1.76 by .22
7	1.76 by .22
8	2.0 by .22
8	2.0 by .22

This test was rather unique, but from the results of tests reported it is evident that this test offers no advantage over that of the simple determination of the weight of charge contained within the detonator.

5. *Ability of detonator to explode similar detonators.*—This test is fully stated in a circular dated September 10, 1903, issued by the chief inspector of explosives (Great Britain) to the manufacturers

---

[a] Photographs of flashes of electric detonators: Trans. Inst. Min. Eng., vol. 15, 1897, p. 203.

and importers of detonators. The detonator is there defined[a] as "A capsule or case of such strength and construction and containing one or the other of the following explosives of the fulminate class in such quantities that the explosion of one capsule or case will communicate the explosion to other capsules or cases: (1) Fulminate of mercury, (2) fulminate of mercury and chlorate of potash, (3) other compositions."

It is obvious from the definition that with this test no discrimination between the detonators of different grades is possible.

6. *Effect on lead block when detonator is fired in bore hole.*—At the Massachusetts Institute of Technology in 1888–89, tests were conducted by Robert C. Williams and J. B. Seager and reported by Frederick W. Clark.[b]

Tests were made of 20 explosives, triple and quintuple detonators (caps) being used. In order that some of the effect of the detonator itself might be eliminated its effect was determined in the following way: The lead block used was a frustum of a cone $5\frac{1}{2}$ inches high, $5\frac{1}{4}$ inches in diameter at the bottom, and 5 inches in diameter at the top. The axial bore was also a frustum of a cone three-fourths of an inch in diameter at the top, five-eighths of an inch in diameter at the bottom, and $2\frac{1}{2}$ inches deep. In casting the blocks the lead was poured when "just barely melted"; the finished block weighed about 45 pounds. The detonator was placed in the bore hole, tamped with dry quartz sand, and fired by means of fuse. As the detonators were slightly less than one-fourth of an inch in diameter the distance between the caps and the walls of the bore hole averaged three-sixteenths of an inch. A tabulation of the results of the tests follows:

Results of firing detonators in bore holes of lead blocks.

Grade of detonator (cap).	Capacity of bore hole.		Difference.	Average.
	Before firing detonator.	After firing detonator.		
	C. c.	*C. c.*	*C. c.*	*C. c.*
"Eagle" triple[a]	14.3	17.0	2.7	
Do	14.3	16.3	2.0	2.3
Do	14.3	16.6	2.3	
"Eagle" quintuple[b]	14.3	17.2	2.9	
Do	14.3	17.5	3.2	3.1

[a] At that time the commercial grade name of the Pittsburgh testing station No. 3 detonator.
[b] At that time the commercial grade name of the Pittsburgh testing station No. 5 detonator.

It is evident that the method of conducting these tests was such that only a part of the energy of the detonator was represented by the expansion of the bore hole because much of the energy was

---

[a] Practical Coal Mining, vol. 2, 1903, p. 237.

[b] Some tests of the relative strength of nitroglycerin and other explosives: Trans. Am. Inst. Min. Eng., vol. 18, 1890, p. 515.

used to disintegrate and pulverize the sand. This was proven by tests made at the Pittsburgh testing station with electric detonators containing similar charges. A No. 3 electric detonator when fired in a cast-lead block with a bore hole of such size that the detonator would fit snugly within it produced an expansion of 5.8 c. c. A similar test with a No. 5 electric detonator gave an expansion of 9.2 c. c.

In the tests at the Massachusetts Institute of Technology, two detonators fired simultaneously within the bore hole produced considerably more than twice the expansion produced by one detonator, probably because the distance between the charge and the sides of the bore hole was less and, accordingly, the charging density was increased. The following tabulated results show this:

Results of firing simultaneously two detonators in bore hole of lead block.

Grade of detonator (cap).	Capacity of bore hole—		Difference.	Average.
	Before firing detonator.	After firing detonator.		
	C. c.	*C. c.*	*C. c.*	*C. c.*
"Eagle" triple..............................	14.3	21.9	7.6	6.8
Do..............................	14.3	20.4	6.1	
"Eagle" quintuple..............................	14.3	20.0	9.7	9.7

Further lead-block tests were made with 13 sensitive explosives, both triple and quintuple detonators (caps) being used. The charge consisted of 6 grams of explosive, loaded and fired as previously described. The conclusion drawn was that explosives when fired with a quintuple detonator produce 9.7 per cent greater expansion than that produced with a triple detonator.

It is evident that in arriving at this conclusion the author did not take into consideration the fact that the quintuple detonator had a charge of 0.80 gram of fulminating composition, that the triple detonator had only a 0.54-gram charge, and that therefore the weight of the total charge, including the quintuple detonator, was increased 4.0 per cent over the weight of the total charge when triple detonators were used. Furthermore, the 4.0 per cent increase in weight represented principally mercury fulminate, a powerful, quick-acting explosive which, under the conditions of the tests, would exert its full effect in enlarging the bore hole. From the data presented, the results can not be properly interpreted as indicating that, with small charges (in this case 6 grams) of an explosive detonating directly under the influence of a detonator, an increase of the force of the explosive was obtained with a detonator of the higher grade.

The results of tests made at the Pittsburgh testing station with sensitive explosives do not substantiate the conclusions drawn. In order to differentiate between grades of electric detonators, it was necessary to use large quantities of insensitive explosives under conditions simulating those of actual blasting operations.

TESTS FOR DETERMINING DIRECTLY THE STRENGTH OF P. T. S. S. ELECTRIC DETONATORS.

CHARACTER OF ELECTRIC DETONATORS TESTED.

Tests for determining directly the strength of electric detonators were made with six grades of P. T. S. S. electric detonators (fig. 1).

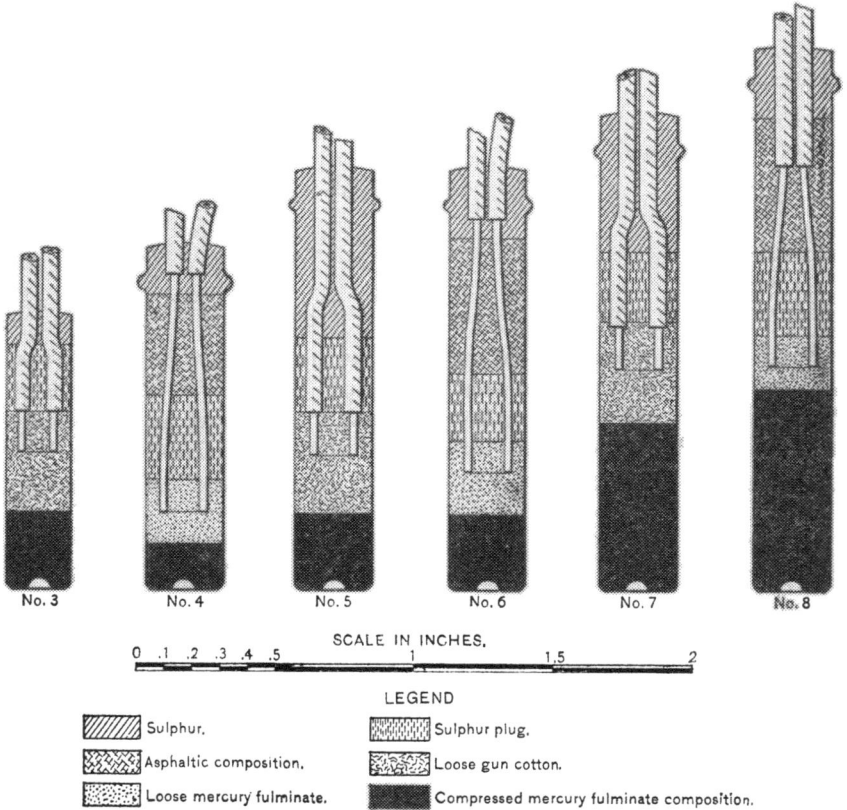

FIGURE 1.—Cross-sectional view of six P. T. S. S. electric detonators.

A physical examination of each showed the results tabulated below. Each measurement represents an average of the measurements of five electric detonators of a given grade.

Results of physical examination of P. T. S. S. electric detonators.

Grade of electric detonator.	Length of shell.	Outside diameter f shell.	Inside diameter of shell.	Thickness of shell.	Length of compressed charge.	Length of priming charge.	Length of sulphur plug.	Length of asphaltic composition, if any.	Length of sulphur filling.
	Inches.	*Inches.*	*Inches.*	*Inches.*	*Inches.*	*Inches.*	*Inches.*	*Inches.*	*Inches.*
No. 3......	1.00	0.234	0.220	0.007	0.28	0.37	0.25	....	0.10
No. 4......	1.25	.274	.260	.007	.16	.24	.31	0.38	.16
No. 5......	1.55	.274	.260	.007	.28	.37	.28	....	.62
No. 6......	1.55	.274	.260	.007	.28	.27	.25	.50	.25
No. 7......	1.75	.274	.260	.007	.62	.38	.25	....	.50
No. 8......	2.00	.274	.260	.007	.75	.20	.31	.50	.24

Details of the wiring of the electric detonators tested are given below:

Details of the wiring of six grades of P. T. S. S. electric detonators.

Grade of electric detonator.	Distance wires projected below sulphur plug.	Distance from end of insulation to end of wires.
	Inches.	*Inches.*
No. 3..	0.16	0.16
No. 4..	.12	.88
No. 5..	.16	.16
No. 6..	.12	.94
No. 7..	.19	.16
No. 8..	.12	.75

The outside diameter and the thickness of the shells were determined with micrometers. The inside diameter of the shells was computed from the figures so determined. For grades Nos. 3, 5, and 7 the priming charge was guncotton. No. 3 electric detonators could not be procured from the manufacturers, so the priming charge, sulphur plug, and sulphur filling were placed in a No. 3 detonator at the Pittsburgh testing station; all other electric detonators were purchased from manufacturers.

The weights and the results of chemical analyses of the charges of the six grades of electric detonators were as follows:

Weights and results of chemical analyses of charges of P. T. S. S. electric detonators.

Grade of electric detonator.	Weight of compressed charge.	Weight of priming charge.	Weight of total charge.	Percentage in compressed charge of—			Percentage in priming charge of—			Percentage in total charge of—		
				Mercury fulminate.	Chlorate of potash.	Guncotton.	Mercury fulminate.	Chlorate of potash.	Guncotton.	Mercury fulminate.	Chlorate of potash.	Guncotton.
	Grams.	*Grams.*	*Grams.*	*Per ct.*	*Per ct.*	*Per ct.*	*Per ct.*	*Per ct.*	*Per ct.*	*Per ct.*	*Per ct.*	*Per ct.*
No. 3 [a].............	0.4920	0.0200	0.5120	87.94	12.06	100.00	....	84.50	11.59	3.91		
No. 4 [a].............	.3255	.3230	.6485	88.51	11.49	....	100.00	94.24	5.76	....		
No. 5 [b].............	.6990	.0240	.7230	89.13	10.87	100.00	....	86.17	10.51	3.32		
No. 6 [b].............	.6485	.3510	.9995	88.82	11.18	....	100.00	92.75	7.25	....		
No. 7 [c].............	1.4854	.0247	1.5101	88.93	11.07	100.00	....	87.47	10.89	1.64		
No. 8 [d].............	1.5110	.3000	1.8110	89.77	10.23	....	100.00	91.47	8.53	....		

[a] Analyst, A. L. Hyde.
[b] Analyst, W. C. Cope.
[c] Analyst, C. A. Taylor.
[d] Analyst, J. H. Hunter.

The results of calorimeter tests are tabulated below:

Results of calorimeter tests of six grades of P. T. S. S. electric detonators.

Grade of electric detonators.	Number of electric detonators used in each test.	Number of tests averaged.	Heat evolved per electric detonator.	Total charge per electric detonator.	Heat evolved per electric detonator on the basis of a charge of 77.7 per cent mercury fulminate and 22.3 per cent chlorate of potash (exact combustion).[a]
			Large calories.	*Grams.*	*Large calories.*
No. 3	30	1	0.35	0.5120	0.36
No. 4	25	2	.48	.6485	.46
No. 5	20	2	.49	.7230	.51
No. 6	15	2	.62	.9995	.71
No. 7	10	2	1.01	1.5101	1.07
No. 8	10	2	1.14	1.8110	1.28

[a] Berthelot, M., Explosives and their power, 1892, p. 470.

The tests were made with the explosives calorimeter[a] of the Pittsburgh testing station and the rise in temperature of the water surrounding the bomb was about 0.140° C., an increase too small to insure the most accurate results. Nevertheless, the results are valuable as showing the potential energy of the electric detonators and that the potential energy is approximately a direct function of the total charge. The last column is added to show how close the heat evolved per electric detonator was to that which was to be expected had the mercury-fulminate composition been of mercury fulminate and chlorate of potash in the proportions necessary for exact combustion.

SQUIRTED LEAD BLOCK TESTS.

Tests of the six grades of electric detonators were made with squirted-lead blocks. The blocks were squirted 2 inches in diameter and were cut 3 inches long. The axial bore hole was drilled a depth equal to the length of the electric detonator to be tested and a diameter such that the electric detonator would fit snugly into it. The volume of the bore hole was measured with water before and after firing the shot. The tendency of the squirted blocks, because of their small diameter (2 inches), to bulge around the sides makes a comparison between the low-grade and the high-grade electric detonator more difficult and makes impossible a comparison of the increase in volume with the weight of total charge. Nevertheless, the volume increases with the weight of total charge as is to be expected.

[a] Hall, Clarence, Snelling, W. O., and Howell, S. P., Investigations of explosives used in coal mines; with a chapter on the natural gas used at Pittsburgh, by G. A. Burrell, and an introduction by C. E. Munroe: Bull. 15, Bureau of Mines, 1912, p. 109.

The results of the tests are tabulated below:

Results of tests P. T. S. S. electric detonators with squirted-lead blocks.

Grade of electric detonator.	Test No.	Volume of bore hole—		Increase of volume after firing electric detonator.	Average increase of volume.	Weight of total charge.
		Before firing detonator.	After firing detonator.			
		C. c.	*C. c.*	*C. c.*	*C. c.*	*Grams.*
No. 3	AA47	0.9	7.5	6.6	6.4	0.5120
	AA48	.9	7.2	6.3		
No. 4	AA45	1.35	11.0	9.6	9.5	.6485
	AA46	1.35	10.8	9.4		
No. 5	AA20	1.8	13.3	11.5	11.3	.7230
	AA27	1.7	12.6	11.1		
No. 6	AA10	1.7	20.0	18.3	18.2	.9995
	AA11	1.7	19.8	18.1		
No. 7	AA18	2.1	38.5	36.4	36.7	1.5101
	AA19	1.9	38.9	37.0		
No. 8	*a* AA16	2.1	49.7	47.6	47.5	1.8110
	a AA17	2.15	49.6	47.45		

a Bottom blown out of block; it was fastened in with paraffin before volume of bore hole was measured.

CAST LEAD BLOCK TESTS.

Tests of the six grades of P. T. S. S. electric detonators were made also with cast-lead blocks. The blocks were cast as solid cylinders 100 mm. in diameter and 100 mm. high. The axial bore hole of each was drilled a depth equal to the length of the electric detonator to be tested, and of a diameter such that the electric detonator would fit snugly into it. The volume of the bore hole was measured with water before and after the shot. When more than two trials were made with any given electric detonator, the two trials that were within 5 per cent variation were selected for averaging. A comparison of the average increase of volume (y) with increase of the weight of total charge (x) shows that the relation $y = 15.5$ ($x = 0.12$) is closely maintained.

Plate I shows the comparative effects of the different electric detonators on the cast-lead blocks.

The details of the cast lead block tests are tabulated below:

Results of tests with cast-lead blocks of P. T. S. S. electric detonators.

Grade of electric detonator.	Test No.	Volume of bore hole—		Increase of volume.	Average increase of volume.	Weight of total charge.	Increase of volume as compared with total charge, by formula $y = 15.5$ ($x = 0.12$).
		Before firing electric detonator.	After firing electric detonator.				
		C. c.	*C. c.*	*C. c.*	*C. c.*	*Grams.*	
No. 3	AA49	0.9	6.6	5.7	5.8	0.5120	6.1
	AA50	.9	6.8	5.9			
No. 4	AA51	1.35	9.3	7.95	7.8	.6485	8.2
	AA52	1.35	9.1	7.75			
No. 5	AA3	1.7	11.1	9.4	9.2	.7230	9.3
	AA39	1.7	10.7	9.0			
No. 6	AA30	1.7	16.0	14.3	14.0	.9995	13.6
	AA55	1.6	15.2	13.6			
No. 7	AA37	1.9	23.0	21.1	21.0	1.5101	21.5
	AA44	1.9	22.8	20.9			
No. 8	AA42	2.1	28.6	26.5	26.2	1.8110	26.2
	AA43	2.1	28.0	25.9			

TESTS BY EXPLOSION OF DETONATING FUSE (CORDEAU DETO-NANT)[a] BY INFLUENCE.

The usual method of firing detonating fuse (cordeau detonant) is to place a detonator on the end of the fuse. Some detonators will explode detonating fuse when not in direct contact with it. Hence, in the expectation that the strength of an electric detonator might be determined by varying the distance between the electric detonator and the detonating fuse, trials with a few electric detonators were made in such a way as to fix for each grade a limiting distance at which no explosion would occur, explosion occurring if the distance were lessened 1 mm.

The detonating fuse was arranged in the four different ways indicated in the following tables:

Results of explosion-by-influence tests in which detonating fuse was placed parallel with electric detonator.

Grade of electric detonator.	Test No.	Trial.	Separating distance.	Result.
			Mm.	
No. 6.......	M243	a	20	No explosion.
		b	10	Do.
		c	5	Do.
		d	0	Do.
		e	0	Do.
		f	0	Do.
		g	0	Do.
		h	0	Do.
No. 8...............	M245	a	0	Do.
		b	0	Do.
		c	0	Do.

Results of explosion-by-influence tests in which side of detonating fuse touched the end of the electric detonator and was at right angles to it.

Grade of electric detonator.	Test No.	Trial.	Result.
No. 6.....................	M244	a	No explosion.
		b	Do.
		c	Do.
		d	Do.
		e	Do.
No. 8.....................	M246	a	Do.
		b	Do.

Results of explosion-by-influence tests in which detonating fuse and electric detonator were placed in the same axial line.

Grade of electric detonator.	Test No.	Trial.	Separating distance.	Result.
			Mm.	
No. 4.....................	M253	a	5	Explosion.
		b	8	No explosion.
		c	6	Do.
		d	6	Do.
		e	6	Explosion.
		f	7	No explosion.
		g	7	Explosion.
		h	8	No explosion.
		i	8	Do.
		j	8	Do.
		k	8	Do.

[a] See p. 66.

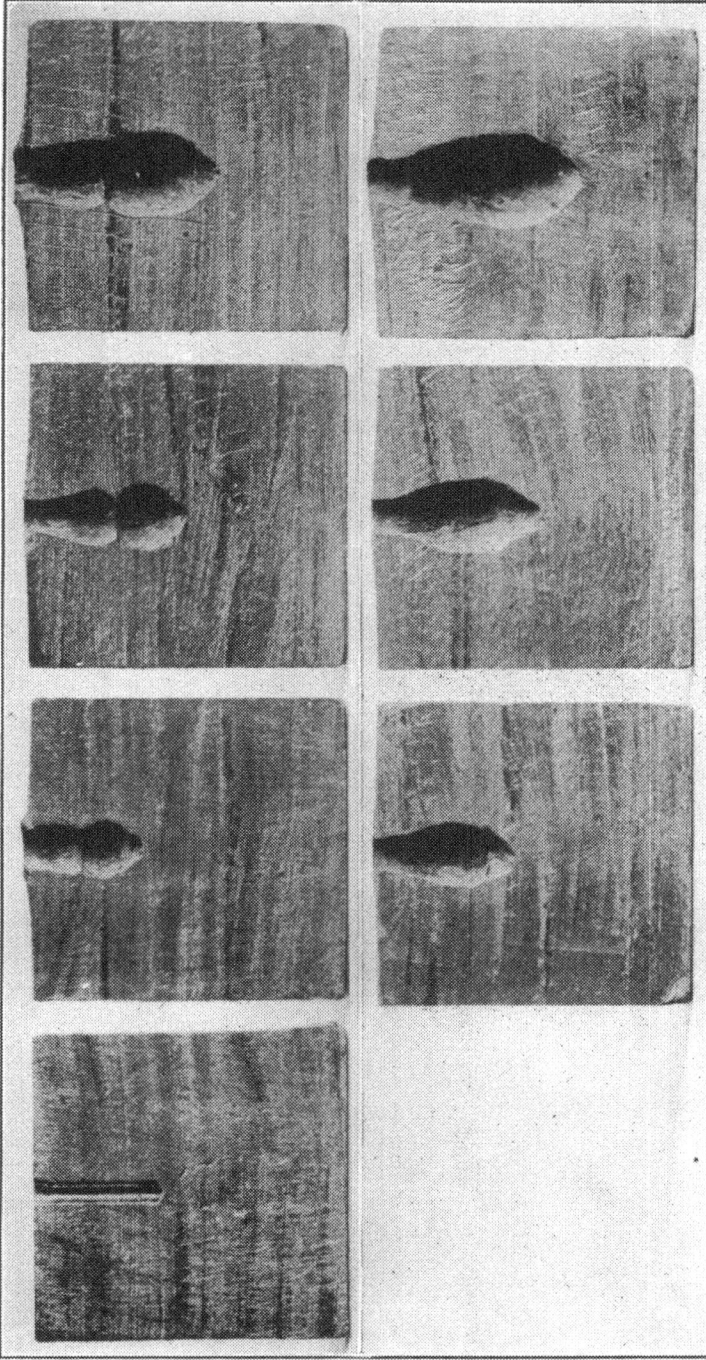

RESULTS OF CAST LEAD BLOCK TESTS OF P. T. S. S. ELECTRIC DETONATORS. UPPER LEFT-HAND CORNER, LEAD BLOCK BEFORE TEST; LEFT TO RIGHT, BLOCKS AS AFFECTED BY DETONATORS NOS. 3, 4, 5, 6, 7, AND 8.

Results of explosion-by-influence tests in which detonating fuse and electric detonator were placed in the same axial line—Continued.

Grade of electric detonator.	Test No.	Trial.	Separating distance.	Result.
			Mm.	
No. 6	M251	a	3	Explosion.
		b	4	Do.
		c	5	Do.
		d	6	Do.
		e	7	No explosion.
		f	7	Do.
		g	7	Do.
		h	7	Explosion.
		i	8	Do.
		j	9	Do.
		k	10	No explosion.
		l	10	Do.
		m	10	Do.
		n	10	Do.
		o	10	Do.
No. 8	M247	a	0	Explosion.
		b	0	Do.
		c	10	No explosion.
		d	5	Do.
		e	1	Explosion.
		f	3	Do.
		g	4	No explosion.
		h	4	Explosion.
		i	5	No explosion.
		j	5	Do.
		k	5	Do.
		l	5	Do.

Results of explosion-by-influence tests in which detonating fuses were placed at right angles to electric detonators but at different distances from them in such a way that axial line of detonating fuse intersected side of electric detonator.

Grade of electric detonator.	Test No.	Trial.	Distance from center line of detonating fuse to end of detonator.	Separating distance.	Result.
			Mm.	*Mm.*	
No. 4	M254	a	5	2	No explosion.
		b	....	1	Explosion.
		c	....	2	No explosion.
		d	....	2	Do.
		e	....	2	Do.
		f	....	2	Do.
No. 6	M252	a	7	5	Do.
		b	....	4	Do.
		c	....	3	Do.
		d	....	2	Do.
		e	....	1	Do.
		f	....	0	Do.
		g	....	0	Explosion.
		h	....	2	No explosion.
		i	....	2	Explosion.
		j	....	3	No explosion.
		k	....	3	Do.
		l	....	3	Do.
		m	....	3	Do.
No. 8	M250	a	10	0	Explosion.
		b	....	4	Do.
		c	....	5	No explosion.
		d	....	5	Do.
		e	....	5	Explosion.
		f	....	6	No explosion.
		g	....	6	Do.
		h	....	6	Do.
		i	....	6	Do.
		j	....	6	Do.

The above results are so much at variance with the established explosive efficiency of detonators (see pp. 45 and 46) that this method of determining the strength of detonators is considered of little value.

The tests made with the No. 4 and the No. 6 electric detonators placed in the same axial line as that of the detonating fuse would indicate that in actual blasting there may be some advantage gained from inserting the electric detonator in the top of the primer or cartridge. Although it has been impossible to show by tests any loss in energy resulting from the detonation of an explosive when the electric detonator is placed in the side of a primer—that is, having the end of the electric detonator intersect the axial line of the primer, it is believed that the former method of insertion is preferable. When the top of the primer is opened and an electric detonator is pushed into it and the paper ends of the cartridge are gathered together and bound with twine, the electric detonator is held firmly in place. When this method is used there is less danger of the wires becoming short-circuited, and it is impossible for the end of the detonator to project through the side of the cartridge, a position that would not only tend to reduce its effectiveness, but would also be a source of danger in loading and tamping the drill hole.

TESTS BY DEPRESSION OF LEAD PLATES.

The strength of the electric detonators was also determined directly by tests involving the depression of lead plates. The details of the tests are indicated in the tabulations following:

Results of depression tests of electric detonators placed on end on $\frac{1}{2}$-inch lead plates.[a]

Grade of electric detonator.	Test No.	Volume of water held in depression.[b]	Diameter of crater.	Depth of crater.	Height of cone on bottom.
		C. c.	*Mm.*	*Mm.*	*Mm.*
No. 3	M264	0.35	11	7	2
No. 4	M262	.50	13	7	4
No. 5	M157	.40	13	6	4
No. 6	M149	.40	13	7	3
No. 7	M151	.40	13	6	2
No. 8	M147	.35	13	7	2

[a] See Pl. II.
[b] The measurement of volume, which was determined with water, was unsatisfactory because of the action of surface tension, and the results are accurate only within 50 per cent. No result obtained agrees even approximately with the established results of the explosive efficiency of electric detonators (see p. 45).

Results of depression tests of electric detonators placed on side on $\frac{1}{2}$-inch lead plates.[a]

MEASURED WITH WATER.

Grade of electric detonator.	Test No.	Volume.[b]	Length of crater.	Width of crater.	Depth of crater.
		C. c.	*Mm.*	*Mm.*	*Mm.*
No. 3	M265	0.30	10	10	4
No. 4	M263	.25	15	10	4
No. 5	M158	.40	15	11	4
No. 6	c M150	.50	21	13	5
No. 7	c M152	.60	20	14	5
No. 8	M148	.90	31	14	6

[a] See Pl. III. [b] See footnote of preceding table.
[c] Bottom of plate slightly raised; not raised in other tests.

SCORING OF LEAD PLATES BY P. T. S. S ELECTRIC DETONATORS NOS. 3, 4, 5, 6, 7, AND 8 PLACED ON END.

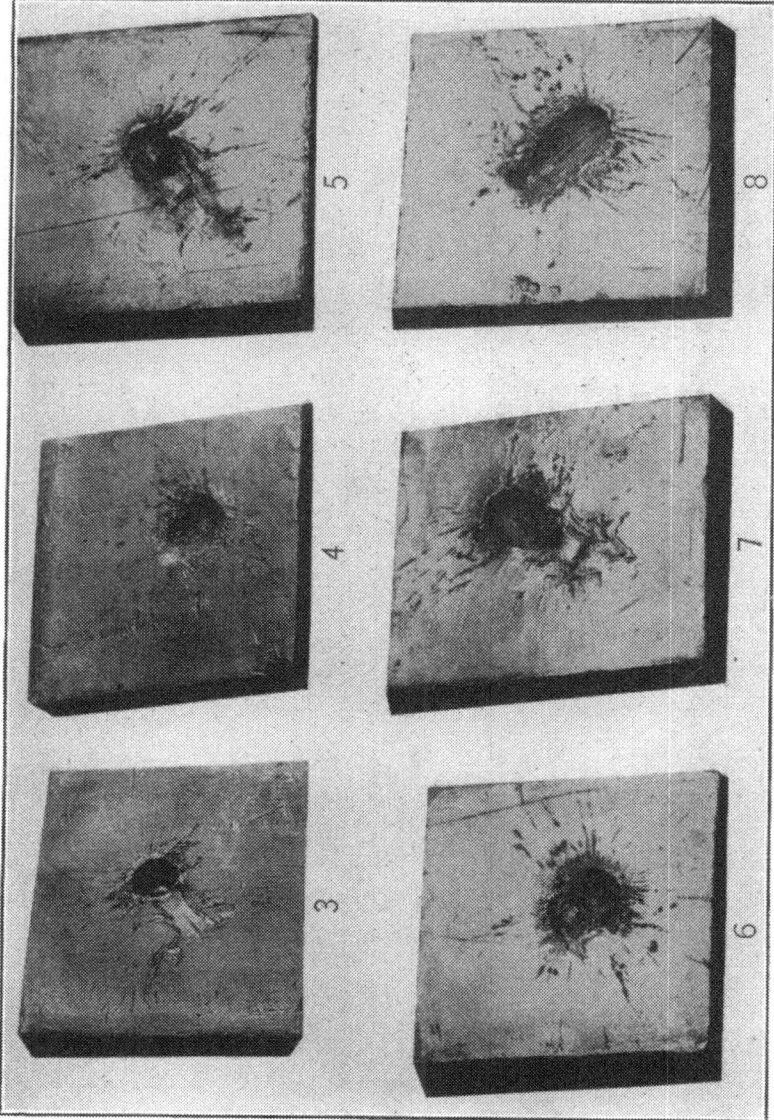

SCORING OF LEAD BLOCKS BY P. T. S. S. ELECTRIC DETONATORS NOS. 3, 4, 5, 6, 7, AND 8 LAID ON SIDE.

Results of depression tests of electric detonators placed on side on ½-inch lead plates—Con.

MEASURED WITH SAND.[a]

Grade of electric detonator.	Test No.	Plate No.	Weight of sand contained in depression No.—					Average of five measurements.	Grand average.	Volume.[b]
			1	2	3	4	5			
			Grams.	Grams.	Grams.	Grams.	Grams.	Grams.	Grams.	C. c.
No. 3.............	M302	1	0.129	0.122	0.121	0.142	0.146	0.132		
		2	.191	.191	.194	.172	.197	.189		
		3	.226	.209	.235	.221	.229	.224	0.182	0.128
No. 4.............	M303	1	.379	.405	.376	.405	.365	.386		
		2	.327	.325	.333	.302	.312	.320		
		3	.415	.393	.397	.404	.386	.399	.368	.259
No. 5.............	M304	1	.361	.359	.340	.355	.381	.359		
		2	.382	.380	.393	.379	.390	.385		
		3	.361	.365	.377	.355	.366	.365	.370	.261
No. 6.............	M301	1	.574	.565	.620	.620	.590	.594		
		2	.580	.586	.607	.569	.596	.588		
		3	.622	.600	.615	.601	.598	.607	.596	.420
No. 7.............	M305	1	.891	.867	.890	.880	.892	.884		
		2	1.002	.994	1.034	.994	1.002	1.005		
		3	1.114	1.076	1.081	1.108	1.144	1.105	.998	.703
No. 8.............	M306	1	1.183	1.173	1.230	1.150	1.244	1.196		
		2	1.245	1.237	1.252	1.272	1.216	1.244		
		3	1.269	1.252	1.257	1.260	1.320	1.272	1.237	.871

[a] The sand was fine and dry.
[b] The volume was computed from the grand average weight by dividing it by the specific gravity of the sand, which was 1.42. This test was acceptable because the volume of the depression varied approximately as the explosive efficiency of the electric detonator.

THE NAIL TEST.

It was evident that the methods previously used for the direct determination of the relative strength of detonators were not satisfactory or accurate. During the latter part of the investigation an endeavor was made to devise a test that would give results approximating those obtained by the indirect tests.

In the tests made with the four No. 6 detonators having different compositions, described later, each electric detonator caused the same amount of energy to be developed from both sensitive and insensitive explosives. However, by the direct methods of testing detonators, one of the No. 6 detonators showed a much higher calorific value than any of the others, and one developed a much greater enlargement of the lead block. Nevertheless it was concluded that although the temperature developed and the volume of gases produced are functions of the efficiency of detonators, the rate of detonation or the rapidity with which the gases are developed is the prime factor and any tests that emphasized this factor should be given consideration.

The test finally decided upon is known as the nail test. This test depends on the angle formed by a nail when a detonator or electric detonator is fired in close proximity to it. For simplicity and cheapness the nail test commends itself.

Four-inch wire finishing nails (20-d.) are used in the test. For the tests herein reported the nails were selected so that they were approximately of the same length, the same gage, and the same weight. The bottom of the electric detonator was placed $1\frac{3}{4}$ inches from the face of the head of the nail and was laid parallel to the nail and separated from it by two 22-gage (0.025-inch) copper wires that were wrapped around the electric detonator. The electric detonator was fastened in position by one strand of a similar copper wire, which was wrapped around it and the nail midway between the ends of the electric detonator. The whole was suspended horizontally in the air in such a manner that the nail was directly above the electric detonator, which was then fired. (Fig. 2.) The impact of the exploding electric detonator bent the nail and projected it upward. Care was taken that the nail was not hurled against any solid surface and further distorted.

Five trials were made with each grade of electric detonator. The angle through which the nail was bent from its normal position was measured. The angle (average of five trials) was taken as a measure of the strength of the electric detonator. The results were as follows:

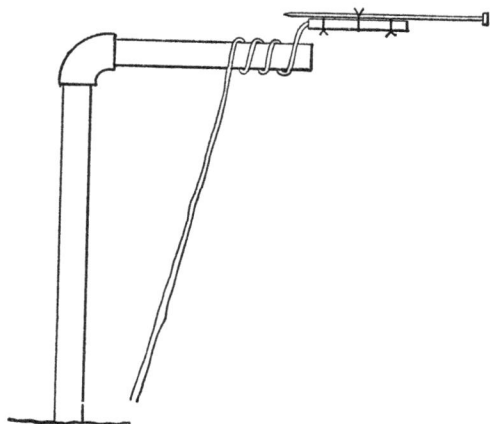

FIGURE 2.—Nail in position for test of electric detonator.

Results of nail tests a of six grades of P. T. S. S. electric detonators.

Grade of electric detonator.	Test No.	Angle of bending resulting from trial No.—					Average.
		1	2	3	4	5	
		°	°	°	°	°	°
No. 3	M279	12	10	8	9	7	9.2
No. 4	M280	12	11	14	11	16	12.8
No. 5	M281	11	13	14	13	8	11.8
No. 6	M288	23	24	25	24	26	24.4
No. 7	M283	60	50	53	54	59	55.2
No. 8	M284	68	78	76	86	98	81.2

a See Pl. IV, A.

The variation in the results of individual trials was largely due to variation in the individual electric detonators. An attempt was made to get more uniform results with annealed nails, but with these there was practically the same variation in results. In such tests, as well as in all physical tests of explosives, discrepancies resulting from unavoidable sources of error can not be eliminated, and, accordingly, only averages should be considered in comparing the practical value of the electric detonators.

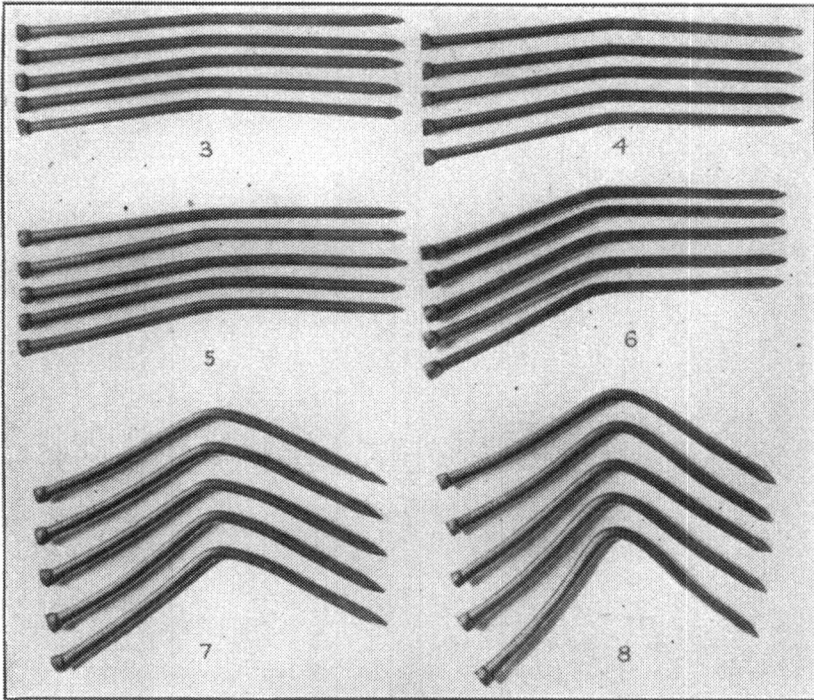

A. RESULTS OF NAIL TESTS OF P. T. S. S ELECTRIC DETONATORS NOS. 3, 4. 5, 6. 7, AND 8.

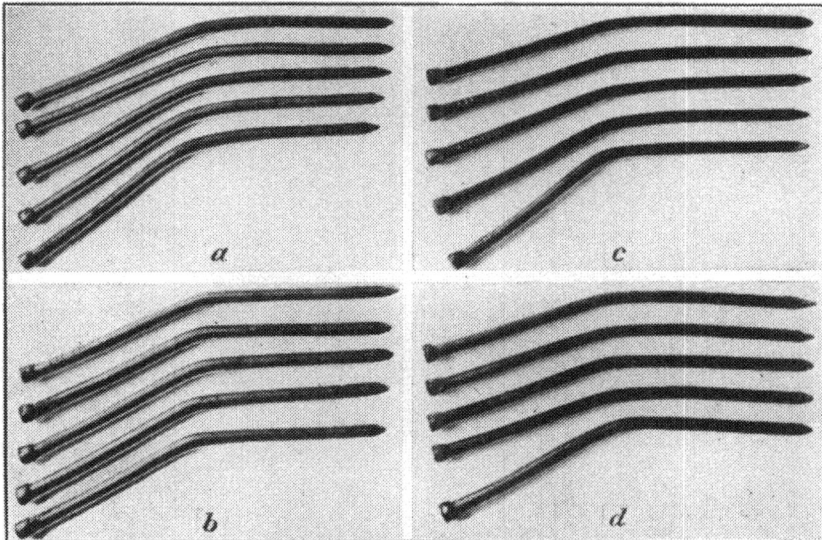

B. RESULTS OF NAIL TESTS OF NO. 6 ELECTRIC DETONATORS. *a*, WESTERN COAST; *b*, SPECIAL; *c*, P. T. S S.; *d*, FOREIGN.

TESTS FOR DETERMINING INDIRECTLY THE STRENGTH OF P. T. S. S. ELECTRIC DETONATORS.

Details of different forms of tests made to determine indirectly the strength of P. T. S. S. electric detonators are given below.

RATE-OF-DETONATION TESTS.[a]

The rate of detonation of the explosive was determined by placing the cartridges end to end in a 28-gage (B. & S.) galvanized-iron tube 42 inches long, which was of slightly larger diameter than the cartridges. The paper ends of each cartridge were cut off squarely in order that the explosive material of the cartridges would be continuous throughout the file, which was a little more than 1 meter long. Four copper wires were inserted through perforations in the tube and the cartridge file so that the distance between adjacent wires made it possible to determine the rate of detonation through the first quarter meter, the second quarter meter, the last half meter, and the entire meter, and the data were so recorded.

Each wire carried an electric current and was attached to a Mettegang recorder in such a way that at the instant the wire was broken a spark was recorded on a rapidly moving soot-covered drum. From the sparks thus recorded and the speed of the drum, the time interval between the breaking of the wires in the meter file was computed and was expressed as rate of detonation in meters per second.

The rate-of-detonation tests were carried on with different explosives as described below.

TESTS WITH AN EXPLOSIVE OF CLASS 1, SUBCLASS a.

In one series of tests sample 1 of an explosive of class 1, subclass a— an ammonium-nitrate explosive containing a sensitizer that is itself an explosive—was used. The cartridges were seven-eighths of an inch in diameter. The results were as follows:

Results of rate-of-detonation tests with sample 1 of an explosive of class 1, subclass a.

Grade of electric detonator.	Test No.	Rate of detonation in tube.							
		First quarter.		Second quarter.		Second half.		Full length.	
		Individual rate.	Average rate.	Individual rate.	Average rate.	Individual rate.	Average rate.	Individual rate.	Average rate.
		Meters per second.	*Meters per second.*	*Meters per second.*	*Meters per second.*	*Meters per second.*	*Meters per second.*	*Meters per second.*	*Meters per second.*
No. 3...	M242	Detonation complete; 16 inches of explosive used.							
	M242	3 inches blown off; 16 inches of explosive used.							
	M248	Detonation complete; 16 inches of explosive used.							
No. ...	M248	Detonation complete; 16 inches of explosive used.							
	M248	Detonation complete; 16 inches of explosive used.							
No. 5...	D963	2,921	....	1,956	20 inches detonated.				
	D980	2 inches blown off.							
No. 6...	D966	2,445	2,458	2,678	2,632	2,205	2,205	2,368	2,362
	D967	2,472		2,586		2,205		2,356	
No. 7...	D964	2,777	2,850	2,586	2,571	2,381	2,374	2,521	2,528
	D965	2,922		2,556		2,368		2,535	
No. 8...	D968	2,184	2,217	2,250	2,285	2,472	2,426	2,337	2,334
	D969	2,250		2,320		2,380		2,331	
Grand average.		....	2,508	....	2,496	....	2,335	....	2,408

[a] For more detailed description of this test, see Bull. 15, Bureau of Mines: Investigations of explosives used in coal mines; with a chapter on the natural gas used at Pittsburgh, by G. A. Burrell, and an introduction by C. E. Munroe, by Clarence Hall, W. O. Snelling, and S. P. Howell, 1912, pp. 92–95.

No detonation occurred in those tests in which 2 or 3 inches of the cartridge was blown off. In those tests in which 16 inches of the explosive was used no attempt was made to determine the rate of detonation.

The grand average indicates that the rate fell off in the last half meter. Individual tests with a given detonator showed remarkable uniformity for each electric detonator of Nos. 6, 7, and 8, and with No. 6 the maximum rate was obtained in the second quarter, with No. 7 in the first quarter, and with No. 8 in the second half. The average rates for the full length of the tube did not vary greatly.

The percentage of complete detonations with each detonator was as follows:

Percentage of complete detonations in rate-of-detonation tests with sample 1 of an explosive of class 1, subclass a.

Grade of electric detonator.	Number of tests.	Number of tests in which incomplete detonation occurred.	Complete detonations.
			Per cent.
No. 3	3	1	67
No. 4	2	0	100
No. 5	2	1	50
No. 6	2	0	100
No. 7	2	0	100
No. 8	2	0	100

These tests show that an explosive of class 1, subclass *a*, that is insensitive, tends more readily to become completely detonated with the higher grades of electric detonators, but that if the explosive detonates at all its rate is independent of the grade of electric detonator used.

The results of tests with sample 2 of an explosive of class 1, subclass *a*, follow. The cartridges used were 1¼ inches in diameter.

Results of rate-of-detonation tests with sample 2 of an explosive of class 1, subclass a.

Grade of electric detonator.	Test No.	Rate of detonation in tube.							
		First quarter.		Second quarter.		Second half.		Full length.	
		Individ-ual rate.	Average rate.	Individ-ual rate.	Average rate.	Individ-ual rate.	Average rate.	Individ-ual rate.	Average rate.
		Meters per second.	*Meters per second.*	*Meters per second.*	*Meters per second.*	*Meters per second.*	*Meters per second.*	*Meters per second.*	*Meters per second.*
No. 3	D1154	3,090		3,423		4,734		3,854	
	D1155	2,973		5,946		2,973		3,398	
	D1156	....	*a* 2,977	5,705	5,529	3,708	3,664	....	*a* 3,668
	D1157	2,884		7,257		3,435		3,750	
	D1158	2,960		5,488		3,516		3,673	
No. 7	D1149	3,836		4,198		3,069		3,477	
	D1150	2,973	3,324	6,286	4,698	2,933	3,340	3,398	3,506
	D1151	2,649		5,174		3,739		3,618	
	D1152	3,836		3,134		3,618		3,532	

a Test No. D1156 not included in average.

The averages for each detonator show a positive acceleration in the second quarter meter and a negative acceleration in the second half meter. The rates for the meter length is within the experimental error, and they were, therefore, practically uniform.

The results of tests with sample 3 of an explosive of class 1, subclass *a*, follow. The cartridges used were 1½ inches in diameter.

Results of rate-of-detonation tests with sample 3 of an explosive of class 1, subclass a.

Grade of electric detonator.	Test No.	Rate of detonation in tube.							
		First quarter.		Second quarter.		Second half.		Full length.	
		Individ-ual rate.	Average rate.	Individ-ual rate.	Average rate.	Individ-ual rate.	Average rate.	Individ-ual rate.	Average rate.
		Meters per second.	*Meters per second.*	*Meters per second.*	*Meters per second.*	*Meters per second.*	*Meters per second.*	*Meters per second.*	*Meters per second.*
No. 3......	D1161	*a* 3,143							
	D1162	4,167	4,308	3,571	3,352	2,980	3,188	3,488	3,510
	D1163	4,450		3,134		3,397		3,532	
No. 7......	D1127	3,477		4,045		3,450		3,589	
	D1128	3,007	3,451	4,363	4,170	3,202	3,327	3,371	3,518
	D1129	3,090		4,541		3,296		3,477	
	D1130	4,231		3,729		3,359		3,636	

a Detonation incomplete; test not averaged.

The average rate for the meter length was practically uniform.

The percentage of complete detonations for each electric detonator was as follows:

Percentage of complete detonations in rate-of-detonation tests with sample 3 of an explosive of class 1, subclass a.

Grade of electric detonator.	Number of tests.	Number of tests in which complete detonation occurred.	Complete detona-tions.
			Per cent.
No. 3..........	3	2	67
No. 7..........	4	4	100

TESTS WITH AN EXPLOSIVE OF CLASS 1, SUBCLASS *b*

The results of tests with sample 1 of an explosive of class 1, subclass *b*—an ammonium-nitrate explosive containing a sensitizer that is not in itself an explosive—follow. The diameter of the cartridges used was 1¾ inches.

Results of rate-of-detonation tests with sample 1 of an explosive of class 1, subclass b.

Grade of electric detonator.	Test No.	Remarks.
No. 3..........	M237	No detonation.
No. 4..........	M237	Do.
No. 5..........	D870	Do.
No. 6..........	D871	Do.
	D869	Do.
	D871	Do.
No. 7..........	D869	Do.
	D871	Do.
No. 8..........	D869	Do.
	D871	Do.

These tests failed to make a discrimination between the different grades of electric detonators, and hence for the purposes of this investigation were useless.

The results of tests with sample 2 of an explosive of class 1, subclass b, follow. The cartridges used were $1\frac{3}{4}$ inches in diameter.

Results of rate-of-detonation tests with sample 2 of an explosive of class 1, subclass b.

Grade of electric detonator.	Test No.	Rate of detonation in tube.							
		First quarter.		Second quarter.		Second half.		Full length.	
		Individual rate.	Average rate.	Individual rate.	Average rate.	Individual rate.	Average rate.	Individual rate.	Average rate.
		Meters per second.	*Meters per second.*	*Meters per second.*	*Meters per second.*	*Meters per second.*	*Meters per second.*	*Meters per second.*	*Meters per second.*
No. 3......	M232	6 inches of cartridge blown off.							
	M241	4 inches of cartridge blown off.							
	M241	4 inches of cartridge blown off.							
	M240	3 inches of cartridge blown off.							
	M240	3 inches of cartridge blown off.							
No. 4......	M231	6 inches of cartridge blown off.							
	M241	4 inches of cartridge blown off.							
	M241	4 inches of cartridge blown off.							
	M240	4 inches of cartridge blown off.							
	M240	4 inches of cartridge blown off.							
No. 5......	M240	4 inches of cartridge blown off.							
	M241	4 inches of cartridge blown off.							
	M241	4 inches of cartridge blown off.							
	D923	4 inches of cartridge blown off.							
	D926	4 inches of cartridge blown off.							
No. 6......	M249	Detonation complete; 16 inches of explosive used.							
	D930	8 inches of cartridge blown off.							
	D931	3,000	3,000	3,750	3,750	3,333	3,333	3,333	3,333
	D932	8 inches of cartridge blown off.							
	D1110	5 inches of cartridge blown off.							
No. 7......	M241	Detonation complete; 16 inches of explosive used.							
	M241	Detonation complete; 16 inches of explosive used.							
	D928	3,358		3,462		3,743		3,475	
	D929	3,462	3,384	3,814	3,589	3,082	3,430	3,333	3,415
	D956	3,333		3,491		3,464		3,437	
No. 8......	M241	Detonation complete; 16 inches of explosive used.							
	D925	2,647		3,214		4,286		3,462	
	D927	3,169		3,750		3,147		3,285	
	D954	3,235	3,072	3,055	3,290	3,358	3,544	3,247	3,319
	D955	3,235		3,142		3,384		3,283	
Grand average.		....	3,180	....	3,460	....	3,475	....	3,357

It is probable that no detonation occurred in those tests in which 3 to 8 inches of the cartridge was blown off. In the trial listed under test M 241 only 16 inches of explosive was used and no attempt was made to determine the rate of detonation.

The grand average shows the tendency of the rate of detonation to increase beyond the first quarter.

It is interesting to observe that the rate of detonation for that 10 centimeters of a $1\frac{3}{4}$-inch cartridge just beyond the electric detonator, as determined with the cordeau detonant, was as follows: For a No. 7 electric detonator, 3,387 meters per second; for a No. 8 electric detonator, 3,387 meters per second.

The percentage of complete detonations for each detonator was as follows:

Percentage of complete detonations in rate-of-detonation tests with sample 2 of an explosive of class 1, subclass b.

Grade of electric detonator.	Number of tests.	Number of tests in which incomplete detonation occurred.	Complete detonations.
			Per cent.
No. 3	5	5	0
No. 4	5	5	0
No. 5	5	5	0
No. 6	5	3	40
No. 7	5	0	100
No. 8	5	0	100

These tests show that an explosive of class 1, subclass *b*, that is insensitive, tends more readily to become completely detonated with the higher grades of electric detonators, but that if the explosive detonates at all its rate is independent of the grade of electric detonator used.

TESTS WITH A 20 PER CENT "STRAIGHT" NITROGLYCERIN DYNAMITE.

The results of tests with a 20 per cent "straight" nitroglycerin dynamite follow. The cartridges were seven-eighths of an inch in diameter.

Results of rate-of-detonation tests with a 20 per cent "straight" nitroglycerin dynamite.

Grade of electric detonator.	Test No.	Rate of detonation in tube.							
		First quarter.		Second quarter.		Second half.		Full length.	
		Individual rate.	Average rate.	Individual rate.	Average rate.	Individual rate.	Average rate.	Individual rate.	Average rate.
		Meters per second.	*Meters per second.*	*Meters per second.*	*Meters per second.*	*Meters per second.*	*Meters per second.*	*Meters per second.*	*Meters per second.*
No. 3	D1096 / D1097	2,528 / 2,781	2,654	3,648 / 2,967	3,308	2,853 / 3,156	3,004	2,918 / 3.007	2,962
No. 4	D1098 / D1099	2,418 / 2,781	2,600	3,423 / 2,967	3,195	2,834 / 2,871	2,852	2,834 / 2,871	2,852
No. 5	D992 / D993	3,225 / 2,747	2,986	2,781 / 2,928	2,854	2,908 / 2,987	2,948	2,947 / 2,908	2,928
No. 6	D1000 / D1001 / D1002	3,729 / *a* 3,034 / 3,947	3,838	2,683 / 3,034 / 2,443	2,563	2,767 / 3,121 / 3,309	3,038	2,933 / 3,077 / 3,158	3,046
No. 7	D1003 / D1004	3,836 / 3,125	3,480	2,500 / 2,586	2,543	2,947 / 3,192	3,070	2,986 / 3,000	2,993
No. 8	D1005 / D1006	3,358 / 3,261	3,310	2,679 / 2,778	2,728	2,980 / 3,041	3,010	2,980 / 3,020	3,000
Grand average.			3,145		2,865		2,987		2,964

a Average rate for the first half meter; rate not included in average.

In the tests where electric detonators Nos. 3, 4, and 5 were used a considerably lower rate of detonation occurred in the first quarter than in the tests where electric detonators Nos. 6, 7, and 8 were used.

Detonation was complete with every grade of electric detonator.

. The figures in the grand average indicate that the rate was influenced by a negative acceleration in the second quarter meter, followed by a positive acceleration in the second half meter, though the contrary was true for electric detonators of grades Nos. 3 and 4. All tests except test D993 conformed to this.

The uniformity of the rates for the last half meter and for the meter for every grade are noteworthy; this uniformity held for individual tests as well as for averages.

TESTS WITH A 40 PER CENT STRENGTH AMMONIA DYNAMITE CONTAINING NITROSUBSTITUTION COMPOUNDS.

The results of tests with a 40 per cent strength ammonia dynamite containing nitrosubstitution compounds follow. The cartridges used were seven-eighths of an inch in diameter and had been repacked.

Results of rate-of-detonation tests with a 40 per cent strength ammonia dynamite containing nitrosubstitution compounds.

Grade of electric detonator.	Test No.	Rate of detonation in tube.							
		First quarter.		Second quarter.		Second half.		Full length.	
		Individual rate	Average rate.	Individual rate.	Average rate.	Individual rate.	Average rate.	Individual rate.	Average rate.
		Meters per second.	*Meters per second.*	*Meters per second.*	*Meters per second.*	*Meters per second.*	*Meters per second.*	*Meters per second.*	*Meters per second.*
No. 5	a D878	....		....		....		2,811	
	D882	2,136		3,929		2,045		2,350	
	D903	2,679	2,498	2,586	3,014	2,284	2,286	2,446	2,412
	D921	2,679		2,528		2,528		2,439	
No. 6	a D875	3,666		2,296		b 2,435		2,659	
	D904	2,368				2,572		2,446	
	a D905	c 2,285	2,363	....	2,594	2,813	2,593	2,521	2,527
	D919	2,472		3,041		2,663		2,695	
	D920	2,250		2,446		2,543		2,439	
No. 7	a D873	2,631		2,273		b 2,668		2,658	
	D906	2,394				2,830		2,557	
	a D907	c 2,367	2,543	....	2,709	2,987	2,619	2,641	2,608
	D918	2,616		2,961		2,528		2,647	
	D959	2,619		2,894		2,500		2,619	
No. 8	a D872	2,716		....		b 2,453		2,514	
	D881	2,558		3,667		2,280		2,597	
	D910	1,542		3,209		3,345		2,561	
	D911	2,123	2,310	2,815	3,146	2,711	2,755	2,557	2,652
	a D915	....		....		....		2,195	
	D916	2,679		3,041		2,744		2,795	
	D917	2,647		3,000		2,695		2,752	

a Rate of detonation not averaged.
b Rate for last three-fourths of a meter.
c Rate for first one-half of a meter.

No tests were made with electric detonators Nos. 3 and 4. The average rate for the meter length increased slightly with the grade of electric detonator used. The fastest rate is recorded for the second quarter meter. The figures for the average rates and for most of the individual tests indicate that the rate increased up to a maximum, and then decreased. With some electric detonators the maximum was reached in the first quarter meter, as in test D903; with others in the second quarter meter, as in test D919; and with others in the second half meter, as in test D910.

If it be assumed that the recorded rate was slightly erratic, but had a general tendency to increase to a maximum, and then to decrease toward an asymptotic normal rate, then the results of all the tests conformed to this assumption.

TESTS WITH A 40 PER CENT STRENGTH AMMONIA DYNAMITE.

The results of tests with a 40 per cent strength ammonia dynamite follow. The cartridges used were $1\frac{1}{4}$ inches in diameter.

Results of rate-of-detonation tests with a 40 per cent strength ammonia dynamite.

Grade of electric detonator.	Test No.	Rate of detonation in tube.							
		First quarter.		Second quarter.		Second half.		Full length.	
		Individual rate.	Average rate.	Individual rate.	Average rate.	Individual rate.	Average rate.	Individual rate.	Average rate.
		Meters per second.	*Meters per second.*	*Meters per second.*	*Meters per second.*	*Meters per second.*	*Meters per second.*	*Meters per second.*	*Meters per second.*
No. 3......	D1139	4,018	3,954	4,412	4,532	4,545	4,242	4,369	4,223
	D1140	4,327		4,592		4,054		4,245	
	D1141	3,516		4,592		4,128		4,054	
No. 8......	D1136	3,437	3,479	4,314	4,868	4,889	4,429	4,293	4,211
	D1137	3,250		5,291		4,417		4,213	
	D1138	3,750		5,000		3,982		4,128	

Only the No. 3 and the No. 8 electric detonators were used. The average rate for the meter length is practically the same for the two detonators. The rate increased to a maximum in the second half meter and then decreased as shown by averages; the results of individual tests confirm this conclusion. The rate in the last half meter corresponded closely with the average rate for the meter length.

TESTS WITH A 35 PER CENT STRENGTH GELATIN DYNAMITE 2 YEARS OLD.

The results of tests with a 35 per cent strength gelatin dynamite (2 years old) follow. The cartridges used were $1\frac{1}{4}$ inches in diameter.

Results of rate-of-detonation tests with a 35 per cent strength gelatin dynamite (2 years old).

Grade of electric detonator.	Test No.	Length of file[a] blown off or detonated.	Percentage inches that detonated.	Average percentage that detonated.	Remarks.
		Inches.	*Per cent.*	*Per cent.*	
No. 3	M231	6.5	15	15.0	Partial detonation.
No. 4	M234	7.5	18	18.5	Do.
No. 5	D887	7.0	0	....	No detonation.
	D888	7.0	0	....	Do.
	D942	13.0	31	....	Partial detonation.
	D943	13.0	31	15.5	Do.
No. 6	D889	7.0	0	....	No detonation.
	D896	18.0	43	....	Partial detonation.
	D958	17.0	40	28.0	Do.
No. 7	D890	12.0	29	....	Do.
	D895	12.0	29	....	Do.
	D957	12.0	29	29.0	Do.
No. 8	D891	18.0	43	....	Do.
	D892	12.0	29	....	Do.
	D940	15.0	36	....	Do.
	D941	14.0	33	35	Do.

a Full length of file, 42 inches.

The evidence of no detonation in tests D887, D888, and D889 was that nothing but the noise of the detonator was audible when the trials were made.

In tests M231 and M234 an 8-inch cartridge was used.

In no trial was more than 18 inches of the 42 inches detonated. The part that detonated, in general, varied directly with the grade of the detonator.

The number of partial detonations with each detonator was as follows:

Number of partial detonations in rate-of-detonation tests with a 35 per cent strength gelatin dynamite (2 years old).

Grade of electric detonator.	Number of tests.	Number of tests in which partial detonation occurred.	Percentage of partial detonations.
			Per cent.
No. 3	1	1	100
No. 4	2	2	100
No. 5	4	2	50
No. 6	3	2	67
No. 7	3	3	100
No. 8	4	4	100

Except with the No. 3 and the No. 4 electric detonators, the number of tests with which was small, the percentage of partial detonations increased with the grade of the electric detonator.

TESTS WITH A 40 PER CENT STRENGTH GELATIN DYNAMITE, FROZEN.

The results of tests with a 40 per cent strength gelatin dynamite (frozen) follow. The diameter of the cartridges used was $1\frac{1}{4}$ inches.

Results of rate-of-detonation tests with a 40 per cent strength gelatin dynamite (frozen).

Grade of electric detonator.	Test No.	Rate of detonation.			
		First quarter.	Second quarter.	Second half.	Total.
		Meters per second.	*Meters per second.*	*Meters per second.*	*Meters per second.*
No. 3 *a*	D 1087	3 inches blown off.			
No. 4	D 1088	4,018	6,429	5,890	5,376
No. 5	D 1089	6,250	7,258	5,769	6,207
No. 6	D 1093	4,167	7,759	6,522	5,921
No. 7	D 1094	4,687	6,429	5,769	5,591
No. 8	D 1095	4,018	7,500	6,522	5,806
Grand average		4,628	7,075	6,094	5,780

a No detonation occurred with the No. 3 electric detonator.

The grand averages show that the maximum rate occurred in the second quarter, with a subsequent falling off in the rate; moreover, each individual test showed similar results, irrespective of the grade of the electric detonator used.

The variation of 14.4 per cent in the average rate is rather high, and is seemingly due to the fact that results with frozen explosives are always erratic.

Complete detonation occurred in each test with each of the six electric detonators except the No. 3, which failed to detonate.

TESTS WITH A 35 PER CENT STRENGTH GELATIN DYNAMITE
3 YEARS OLD.

The results of tests with a 35 per cent strength gelatin dynamite (three years old) follow. The cartridges used were $1\frac{1}{2}$ inches in diameter.

Results of rate-of-detonation tests with 35 per cent strength gelatin dynamite (3 years old).

Grade of electric detonator.	Test No.	Remarks.
No. 3	M238	No detonation.
No. 4	M238	Do.
No. 5	D868	Do.
No. 7	D866	Do.
No. 8	D865	Do.

The explosive was so old and insensitive to detonation that for the purpose of discriminating between grades of electric detonators it was useless, because in no test did detonation occur. No tests were made with the No. 6 electric detonator.

SMALL LEAD BLOCK TESTS.[a]

The lead blocks used in the small lead block tests were squirted with a diameter of 1½ inches and were cut to a length of 2½ inches. An annealed steel disk 1½ inches in diameter and one-quarter inch high was placed above each block and above this was placed the 100-gram charge of the explosive, held in position by a paper sleeve wrapped around the block and the disk and extending above them. The electric detonator used was centrally placed in the top of the charge. When the explosion was fired, the block rested on a firm horizontal steel base. The compression of the block was determined by measuring the difference in the height of the block before and after firing.

TESTS WITH A 20 PER CENT "STRAIGHT" NITROGLYCERIN DYNAMITE WITH 6 PER CENT OF ADDED WATER.

The results of tests of a 20 per cent "straight" nitroglycerin dynamite follow. The explosive contained 6 per cent of added water:

Results of small lead block tests with a 20 per cent "straight" nitroglycerin dynamite containing 6 per cent of added water.

Grade of electric detonator.	Test No.	Compression.	Average compression.
		Mm.	*Mm.*
No. 3	B755	14.00	
	B764	14.25	14.08
	B773	14.00	
No. 4	B756	15.00	
	B765	14.50	14.67
	B774	14.50	
No. 5	B757	14.25	
	B766	14.00	14.08
	B775	14.00	
No. 6	B761	15.00	
	B770	15.00	15.00
	B779	15.00	
No. 7	B762	15.25	
	B771	14.75	15.25
	B780	15.75	
No. 8	B763	15.50	
	B772	15.75	15.50
	B781	15.25	

The No. 8 electric detonator produced a compression 9.6 per cent greater than that of the No. 3 electric detonator; in general with the explosive tested, the compression increased with the grade of the detonator. The No. 4 electric detonator, however, developed more energy than did the No. 5.

---

[a] For a more extended description of the small lead block test, see Bull. 15, Bureau of Mines: Investigations of explosives used in coal mines; with a chapter on the natural gas used at Pittsburgh, by G. A. Burrell, and an introduction by C. E. Monroe, by Clarence Hall, W. O. Snelling, and S. P. Howell, 1912, pp. 113-114.

TESTS WITH A 20 PER CENT "STRAIGHT" NITROGLYCERIN DYNAMITE, FROZEN AND CONTAINING LESS THAN 6 PER CENT OF ADDED WATER.

The results of tests with a 20 per cent "straight" nitroglycerin dynamite (frozen and containing no added water or 2.5 or 4 per cent of added water) are tabulated below:

Results of small lead block tests of a 20 per cent "straight" nitroglycerin dynamite (frozen and containing less than 6 per cent of added water).

Grade of electric detonator.	Test No.	Temperature of frozen explosive.	Percentage of added water.	Compression.	Average compression.
		° C.		Mm.	Mm.
No. 3	B616	+2.0	0	14.25	
	B622	−1.0	2.5	13.25	13.33
	B647	−9.0	4.0	12.50	
No. 4	B617	+2.0	0	14.50	
	B623	−1.0	2.5	13.25	13.42
	B648	−9.0	4.0	12.50	
No. 5	B618	+2.0	0	13.50	
	B624	−1.0	2.5	13.00	13.00
	B649	−9.0	4.0	12.50	
No. 6	B619	+2.0	0	13.50	
	B625	−1.0	2.5	13.25	13.25
	B653	−9.0	4.0	13.00	
No. 7	B620	+2.0	0	15.00	
	B625	−1.0	2.5	13.50	13.83
	B654	−9.0	4.0	13.00	
No. 8	B621	+2.0	0	15.25	
	B627	−1.0	2.5	13.25	13.92
	B655	−9.0	4.0	13.25	

The tests showed the tendency of the electric detonators to increase slightly in explosive efficiency with the grade, but again the No. 3 and the No. 4 electric detonators showed an increase over the No. 5 and even over the No. 6.

TESTS WITH A 20 PER CENT "STRAIGHT" NITROGLYCERIN DYNAMITE, FROZEN AND CONTAINING 6 PER CENT OF ADDED WATER.

As no failures had occurred with any of the electric detonators, when tested with the 20 per cent "straight" nitroglycerin dynamite, a sample of that explosive with 6 per cent of added water was frozen (temperature 9° C.) and was tested, with results as follows:

Results of small lead block tests with a 20 per cent "straight" nitroglycerin dynamite (frozen and containing 6 per cent of added water).

Grade of electric detonator.	Test No.	Compression.	Average compression.
		Mm.	Mm.
No. 3	B728	a 0.00	
	B737	a.00	0.00
	B746	a.00	
No. 4	B729	a.00	
	B738	a.00	.00
	B747	a.00	

a Incomplete detonation.

Results of small lead block tests with a 20 per cent "straight" nitroglycerin dynamite (frozen and containing 6 per cent of added water)—Continued.

Grade of electric detonator.	Test No.	Compression.	Average compression.
		Mm.	*Mm.*
No. 5	B730	*a* 0.00	
	B739	*a*.00	0.00
	B748	*a*.00	
No. 6	B734	*a*.00	
	B743	*a*.00	.00
	B752	*a*.00	
No. 7	B735	12.75	
	B744	13.75	9.00
	B753	*a*.50	
No. 8	B736	12.75	
	B745	13.75	9.17
	B754	*a* 1.00	

a Incomplete detonation.

The number of complete detonations with each detonator was as follows:

Number of complete detonations with a 20 per cent "straight" nitroglycerin dynamite (frozen and containing 6 per cent of added water).

Grade of electric detonator.	Number of tests.	Number of tests in which complete detonation occurred.	Percentage of complete detonations.
No. 3	3	0	0
No. 4	3	0	0
No. 5	3	0	0
No. 6	3	0	0
No. 7	3	2	67
No. 8	3	2	67

The explosive was very insensitive and complete detonation occured only with the No. 7 and No. 8 electric detonators and with them in only two out of three trials with each.

TESTS WITH A 40 PER CENT STRENGTH AMMONIA DYNAMITE WITH 6 PER CENT OF ADDED WATER.

The results of tests with a 40 per cent strength ammonia dynamite with 6 per cent of added water are tabulated below:

Results of small lead block tests with a 40 per cent strength ammonia dynamite containing 6 per cent of added water.

Grade of electric detonator.	Test No.	Compression.	Average compression.
		Mm.	*Mm.*
No. 3	B656	7.25	
	B665	8.50	
	B674	9.50	8.25
	B683	8.25	
	B692	7.75	

Results of small lead block tests with a 40 per cent strength ammonia dynamite containing 6 per cent of added water—Continued.

Grade of electric detonator.	Test No.	Compression.	Average compression.
		Mm.	*Mm.*
No. 4	B657	7.25	
	B666	8.75	
	B675	8.50	8.20
	B684	8.00	
	B693	8.50	
No. 5	B658	6.00	
	B667	7.25	
	B676	8.75	7.70
	B685	8.00	
	B694	8.50	
No. 6	B662	9.75	
	B671	8.75	
	B680	9.25	8.95
	B689	7.75	
	B698	9.25	
No. 7	B663	8.00	
	B672	8.75	
	B681	10.00	9.15
	B690	10.25	
	B699	8.75	
No. 8	B664	8.75	
	B673	9.50	
	B682	10.75	9.70
	B691	9.75	
	B700	9.75	

This explosive showed a marked tendency to be erratic both with the higher and with the lower grades of electric detonators.

The explosive efficiency of the electric detonators increased with the grade of the electric detonator, except that the efficiency of the No. 5 electric detonator was considerably low and that of the No. 3 a trifle high.

TESTS WITH A 40 PER CENT STRENGTH GELATIN DYNAMITE, FROZEN.

Following are the results (Pl. V, *A*) of tests with a 40 per cent strength gelatin dynamite that was in a frozen condition:

Results of small lead block tests with a 40 per cent strength gelatin dynamite (frozen).

Grade of electric detonator.	Test No.	Temperature of frozen explosive.	Compression.	Average compression.
		° C.	*Mm.*	*Mm.*
No. 3	B633	−2.5	*a* 3.00	
	B638	−5.0	16.75	
	B701	+0.5	13.00	14.25
	B710	+2.5	13.50	
	B719	+2.5	13.75	
No. 4	B629	−4.5	*a* 1.50	
	B639	−5.0	15.50	
	B702	+0.5	10.75	12.62
	B711	+2.5	11.00	
	B720	+2.5	13.25	
No. 5	B630	−4.5	*a* 1.00	
	B640	−5.0	17.25	
	B703	+0.5	13.75	13.12
	B712	+2.5	10.75	
	B721	+2.5	10.75	

a Incomplete detonation.

Results of small lead block tests with a 40 per cent strength gelatin dynamite (frozen)—Con.

Grade of electric detonator.	Test No.	Temperature of frozen explosive.	Compression.	Average compression.
		° C.	Mm.	Mm.
No. 6	B631	−4.5	12.50	
	B644	−5.0	19.25	
	B707	+0.5	14.25	14.95
	B716	+2.5	14.25	
	B725	+2.5	14.50	
No. 7	B632	−4.5	14.50	
	B645	−5.0	20.25	
	B708	+0.5	18.00	17.00
	B717	+2.5	16.25	
	B726	+2.5	16.00	
No. 8	B637	−2.5	15.75	
	B646	−5.0	18.00	
	B709	+0.5	19.75	17.80
	B718	+2.5	17.25	
	B727	+2.5	18.25	

The results were very erratic. The strength of the detonators increased with the grade of the electric detonator used, as shown by the average compression, except that the compression with the No. 3 electric detonator was comparatively high.

The number of complete detonations with each detonator was as follows:

Number of complete detonations in small lead block tests with a 40 per cent strength gelatin dynamite (frozen).

Grade of electric detonator.	Number of tests.	Number of tests in which complete detonation occurred.	Percentage of complete detonations.
			Per cent.
No. 3	5	4	80
No. 4	5	4	80
No. 5	5	4	80
No. 6	5	5	100
No. 7	5	5	100
No. 8	5	5	100

The results tabulated above indicate that the tendency to complete detonation increases with the grade of the electric detonator used.

EXPLOSION-BY-INFLUENCE TESTS.[a]

Explosion-by-influence tests were conducted by placing two cartridges of an explosive at a definite distance apart; each cartridge was in a vertical position, one being directly above the other. The electric detonator was placed in the lower end of the lower cartridge, so that the lower cartridge on detonation either did or did not cause

---

a For a more extended description of the test, see Bull. 15, Bureau of Mines: Investigations of explosives used in coal mines; with a chapter on the natural gas used at Pittsburgh, by G. A. Burrell, and an introduction by C. E. Munroe, by Clarence Hall, W. O. Snelling, and S. P. Howell, 1912, p. 100.

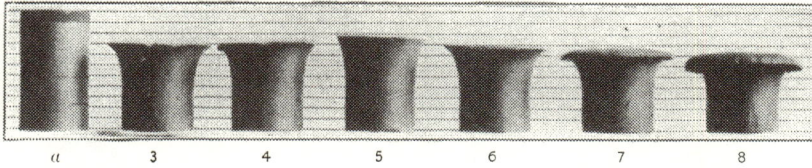

A. RESULTS OF SMALL LEAD BLOCK TESTS OF P. T. S. S. ELECTRIC DETONATORS NOS. 3, 4, 5. 6, 7, AND 8. *a*, BLOCK BEFORE TEST.

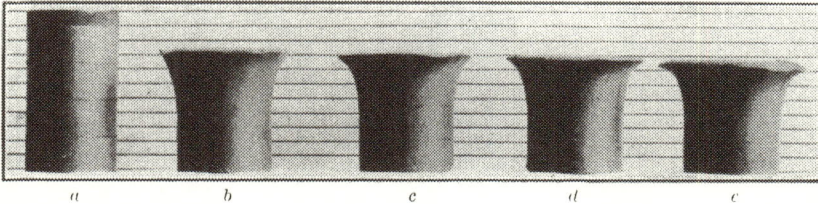

B. RESULTS OF SMALL LEAD BLOCK TESTS OF NO. 6 ELECTRIC DETONATORS. *a*, BLOCK BEFORE TEST; *b*, WESTERN COAST; *c*, SPECIAL; *d*, P. T. S. S.; *e*, FOREIGN.

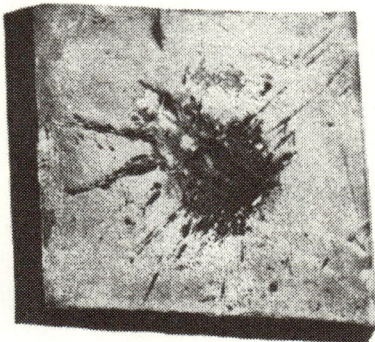

WESTERN COAST.

SPECIAL.

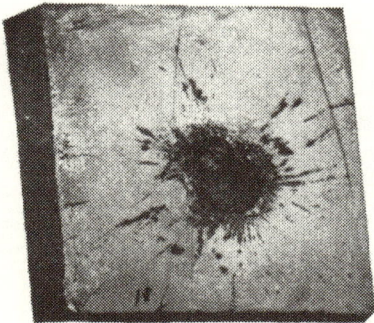

P. T. S. S.

FOREIGN.

C. SCORING OF LEAD PLATES BY FOUR NO. 6 ELECTRIC DETONATORS LAID ON SIDE.

detonation of the upper cartridge. The separating distance, established by successive trials, was but 1 inch greater than that at which the upper cartridge would detonate. With one explosive, however, certain trials were run with the cartridges separated by a given distance, and the number of times that the upper cartridge did or did not detonate was recorded.

TESTS WITH AN EXPLOSIVE OF CLASS 1, SUBCLASS *a*.

The results of tests with an explosive of class 1, subclass *a* (an ammonium-nitrate explosive containing a sensitizer that is itself an explosive), are tabulated below. The average weight of the cartridges was 166 grams and they measured $1\frac{1}{4}$ by 8 inches.

Results of explosion-by-influence tests with an explosive of class 1, subclass a.

Grade of electric detonator.	Test No.	Distance separating cartridges.	Result on upper cartridge.	Distance established at—
		Inches.		*Inches.*
No. 3....	J874	3	Did not explode...	
	J875	2	Exploded.........	
	J876	3	.....do...........	4
	J877	4	Did not explode...	
	J878	4	.....do...........	
No. 4....	J870	2	Exploded.........	
	J871	3	.....do...........	4
	J872	4	Did not explode...	
	J873	4	.....do...........	
No. 5....	J764	2	.....do...........	
	J765	1	Exploded.........	2
	1766	2	Did not explode...	
No. 6....	J741	4	.....do...........	
	J742	3	.....do...........	
	J743	2	.....do...........	3
	J744	1	Exploded.........	
	J745	2	.....do...........	
	J746	3	Did not explode...	
No. 7....	J747	3	.....do...........	
	J748	2	Exploded.........	3
	J749	3	Did not explode...	
No. 8....	J750	3	.....do...........	
	J751	2	Exploded.........	3
	J752	3	Did not explode...	

These tests did not discriminate as to the relative efficiency of the different grades of electric detonators; the efficiency of the low-grade electric detonators was at least as great as that of the high-grade electric detonators.

TESTS WITH AN EXPLOSIVE OF CLASS 4.

Following are the results of tests with an explosive of class 4 (an explosive in which the characteristic material is nitroglycerin). Except for the trials under test J896, the average weight of each cartridge was 161 grams and the size of each was $1\frac{1}{4}$ by 8 inches. In the trials under test J896 the lower cartridge weighed 161 grams and the upper one weighed 110 grams, being only 5 inches long. In all of the tests in which the distance separating cartridges was 5 inches, the bottoms of the cartridges (as packed) faced each other, whereas in all of the tests in which the separating distance was 4 inches, the tops of the cartridges faced each other.

Results of explosion-by-influence tests with an explosive of class 4.

Grade of electric detonator.	Test No.	Distance separating cartridges.	Result—upper cartridge.
		Inches.	
No. 3	J895	5	Did not explode.
	J895	5	Do.
	J895	5	Exploded.
	J895	5	Did not explode.
	J896	4	Do.
	J896	4	Do.
	J896	4	Do.
No. 4	J895	5	Do.
	J895	5	Do.
	J895	5	Do.
	J895	5	Exploded.
	J896	4	Do.
	J896	4	Do.
	J896	4	Did not explode.
No. 5	J895	5	Do.
	J895	5	Do.
	J895	5	Do.
	J895	5	Exploded.
	J896	4	Did not explode.
	J896	4	Do.
	J896	4	Do.
No. 6	J895	5	Do.
	J895	5	Do.
	J895	5	Do.
	J895	5	Exploded.
	J896	4	Do.
	J896	4	Do.
	J896	4	Did not explode.
No. 7	J895	5	Do.
	J895	5	Do.
	J895	5	Do.
	J895	5	Exploded.
	J896	4	Do.
	J896	4	Do.
	J896	4	Did not explode.
No. 8	J895	5	Do.
	J895	5	Do.
	J895	5	Do.
	J895	5	Do.
	J896	4	Do.
	J896	4	Exploded.
	J896	4	Do.

The following tabulation shows the number of explosions of the upper cartridge:

Percentage of explosions of the upper cartridge in explosion-by-influence tests with an explosive of class 4.

Grade of electric detonator.	Number of tests.	Number of explosions of the second cartridge.	Percentage of explosions.
No. 3	7	1	14
No. 4	7	3	43
No. 5	7	1	14
No. 6	7	3	43
No. 7	7	3	43
No. 8	7	2	29

TESTS WITH A 40 PER CENT STRENGTH AMMONIA DYNAMITE CONTAINING NITROSUBSTITUTION COMPOUNDS.

Following are tabulated the results of tests with a 40 per cent strength ammonia dynamite containing nitrosubstitution compounds. The cartridges used were 1¼ by 8 inches, their average weight being 226 grams.

Results of explosion-by-influence tests with a 40 per cent strength ammonia dynamite containing nitrosubstitution compounds.

Grade of electric detonator.	Test No.	Distance separating cartridges.	Result on upper cartridge.	Distance established at—
		Inches.		*Inches.*
No. 5..	J708	8	Exploded.........	
	J709	9	Did not explode...	9
	J710	9	.....do...........	
No. 6..	J689	14	.....do...........	
	J690	12	.....do...........	
	J691	9	.....do...........	
	J692	7	.....do...........	
	J693	6	.....do...........	
	J694	4	Exploded.........	8
	J695	5	.....do...........	
	J696	6	.....do...........	
	J697	7	.....do...........	
	J698	8	Did not explode...	
	J699	8	.....do...........	
No. 7....,.	J720	9	.....do...........	
	J721	8	Exploded.........	9
	J722	9	Did not explode...	
No. 8........	J704	8	Exploded.........	
	J705	9	.....do.....	10
	J706	10	Did not explode...	
	J707	10	.....do...........	

No tests made with detonators Nos. 3 and 4.

TESTS WITH A 35 PER CENT STRENGTH GELATIN DYNAMITE 2 YEARS OLD.

Following are the results of tests with a 35 per cent strength gelatin dynamite (two years old). The average weight of each cartridge was 265 grams and the size of each 1¼ by 8 inches.

Results of explosion-by-influence tests with a 35 per cent strength gelatin dynamite (2 years old).

Grade of electric detonator.	Test No.	Distance separating cartridges.	Result on upper cartridge.
		Inches.	
No. 6..	J724	6	Did not explode.
	J725	5	Do.
	J726	4	Do.
	J727	2	Do.
	J728	0	Do.
	J729	0	Do.
No. 7..	J730	0	Do.
	J731	0	Do.
No. 8..	J732	0	Do.
	J733	0	Do.

No tests were made with the No. 3, the No. 4, or the No. 5 electric detonators.

The tests failed to discriminate between the different grades of electric detonators, except to the limited extent that in two trials the lower cartridge failed to detonate completely once with the No. 6. In no trial did the detonation of the lower cartridge cause the detonation of the upper cartridge.

PERCENTAGES OF DETONATIONS IN INDIRECT TESTS OF P. T. S. S. ELECTRIC DETONATORS.

The percentages of detonations in the indirect tests of the P. T. S. S. electric detonators are given below. The percentages of detonations in the tests of each electric detonator are also averaged, each average percentage having a value proportional to the number of tests from which computed; that is, each percentage is multiplied by the number of tests it represents, and the sum of the products is divided by the total number of tests of the electric detonator considered.

Percentages of detonations in indirect tests of P. T. S. S. electric detonators.

Class of explosive.	Kind of test.	Grade of electric detonator.											
		No. 3.		No. 4.		No. 5.		No. 6.		No. 7.		No. 8.	
		Percentage of detonations.	Number of tests.	Percentage of detonations.	Number of tests.	Percentage of detonations.	Number of tests.	Percentage of detonations.	Number of tests.	Percentage of detonations.	Number of tests.	Percentage of detonations.	Number of tests.
		Per cent.		*Per cent.*		*Per cent.*		*Per cent.*		*Per cent.*		*Per cent.*	
Class 1, subclass *b*	Rate of detonation.	0	5	0	5	0	5	40	5	100	5	100	5
Class 1, subclass *a*	....do......	67	3	100	2	50	2	100	2	100	2	100	2
40 per cent strength gelatin dynamite (frozen).	....do......	0	1	100	1	100	1	100	1	100	1	100	1
35 per cent strength gelatin dynamite (two years old).	....do......	100	1	100	2	50	4	67	3	100	3	100	4
20 per cent "straight" nitroglycerin dynamite (containing 6 per cent of added water and frozen).	Small lead block.	0	3	0	3	0	3	0	3	67	3	67	3
40 per cent strength gelatin dynamite (frozen).	....do......	80	5	80	5	80	5	100	5	100	5	100	5
Total number of tests			18		18		20		19		19		20
Average percentage of detonations.		38.9		50.0		40.0		63.2		94.7		95.0	

COMPARATIVE EXPLOSIVE EFFICIENCY.

The percentages of explosive efficiency of the different types of P. T. S. S. electric detonators were obtained by averaging all tests in which the rate of detonation or compression was determined for all the electric detonators. The percentages of the individual electric detonators were also averaged, each average percentage having a value proportional to the number of tests from which computed; that is, each percentage is multiplied by the number of tests it represents, and the sum of the products is divided by the total number of tests of the electric detonator considered. In each case the percentage of explosive efficiency of the No. 6 electric detonator is assigned a value of 100 and is used as the unit of comparison.

Explosive efficiency of P. T. S. S. electric detonators.

Class and grade of explosive.	Kind of test.	No. 3.				No. 4.				No. 5.				No. 6.				No. 7.				No. 8.			
		Rate (meters per second).	Compression.	Percentage of explosive efficiency.	Number of tests.	Rate (meters per second).	Compression.	Percentage of explosive efficiency.	Number of tests.	Rate (meters per second).	Compression.	Percentage of explosive efficiency.	Number of tests.	Rate (meters per second).	Compression.	Percentage of explosive efficiency.	Number of tests.	Rate (meters per second).	Compression.	Percentage of explosive efficiency.	Number of tests.	Rate (meters per second).	Compression.	Percentage of explosive efficiency.	Number of tests.
			Mm.	P. ct.			Mm.	P. ct.			Mm.	P. ct.			Mm.	P. ct.			Mm.	P. ct.			Mm.	P. ct.	
20 per cent "straight" nitroglycerin dynamite in 1¼-inch cartridges.	Rate of detonation.	2,962		97.2	2	2,852		93.6	2	2,928		96.1	2	3,046		100.0	3	2,993		98.3	2	3,000		98.5	2
35 per cent strength gelatin dynamite 2 years old in 1¼-inch cartridges.	...do....	a 6.5		37.1	1	a 7.5		42.9	1	a 13.0		74.3	1	a 17.5		100.0	2	a 12.0		68.6	3	a 14.8		84.6	4
20 per cent "straight" nitroglycerin dynamite (containing 6 per cent of added water).	Small lead block.		14.08	93.9	3		14.67	97.8	3		14.08	93.9	3		15.00	100.0	3		15.25	101.7	3		15.50	103.3	3
20 per cent "straight" nitroglycerin dynamite (containing 0 to 4 per cent of added water and frozen).	...do....		13.33	100.6	3		13.42	101.3	3		13.00	98.1	3		13.25	100.0	3		13.83	104.4	3		13.92	105.1	3
40 per cent strength gelatin dynamite (frozen).	...do....		14.25	95.3	4		12.62	81.7	4		13.12	87.8	4		14.95	100.0	5		17.00	113.7	5		17.80	119.1	5
40 per cent strength ammonia dynamite (containing 6 per cent of added water).	...do....		8.25	92.2	5		8.20	91.6	5		7.70	86.0	5		8.95	100.0	5		9.15	102.2	5		9.70	108.4	5
Average... Total number of tests.				92.1	18			89.6	18			89.4	19			100.0	21			100.0	21			104.5	22

a Inches that detonated.

Average explosive efficiency of electric detonators Nos. 3, 4, and 5 in 55 tests, 90.4.
Average explosive efficiency of electric detonators Nos. 6, 7, and 8 in 64 tests, 101.6.

COMPARATIVE EXPLOSIVE EFFICIENCY OF P. T. S. S. ELECTRIC DETONATORS.

The tabulation below shows the comparative explosive efficiency (fig. 3) of the six grades of P. T. S. S. electric detonators:

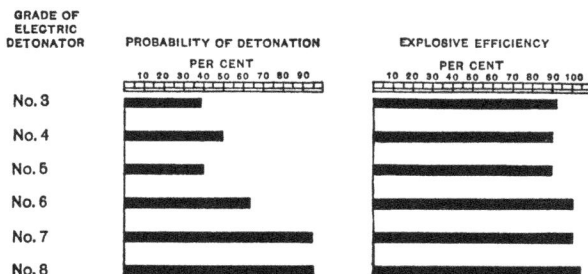

FIGURE 3.—Comparative explosive efficiency of six grades of P. T. S. S. electric detonators as determined by indirect tests.

Comparative explosive efficiency of six grades of P. T. S. S. electric detonators.

Grade of electric detonator.	Probability of detonation.	Explosive efficiency for those tests in which detonation occurred.
	Per cent.	*Per cent.*
No. 3	38. 9	92. 1
No. 4	50. 0	89. 6
No. 5	40. 0	89. 4
No. 6	63. 2	100. 0
No. 7	94. 7	100. 0
No. 8	95. 0	104. 5

TESTS OF FOUR NO. 6 ELECTRIC DETONATORS OF DIFFERENT MAKES.

In the tests of P. T. S. S. electric detonators as described above the composition of the fulminating charge was practically the same in each electric detonator, although there was variation in the weight of the charge. In the tests, the results of which are tabulated below, four No. 6 electric detonators manufactured by different makers were used. The weight of charge of each of the No. 6 electric detonators tested was approximately 1 gram, but each electric detonator had a different composition. The electric detonators were representative of all electric detonators used in the United States, and the tests made are of especial importance for the reason that they established for each electric detonator the charge equivalent to the Pittsburgh testing station standard electric detonators.

PHYSICAL EXAMINATION.

A physical examination was made of each of the four electric detonators (fig. 4), the results being given in the following tabulation. Each item represents an average of measurement of five electric detonators.

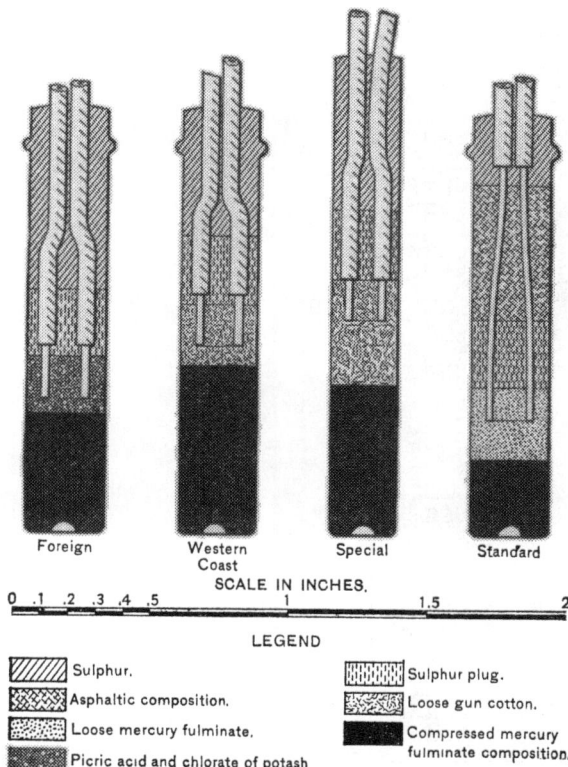

FIGURE 4.—Cross-sectional view of four No. 6 electric detonators of different makes.

Results of physical examination of four No. 6 electric detonators of different makes.

Kind of electric detonator.	Length of shell.	Outside diameter of shell.	Inside diameter of shell.	Thickness of shell.	Length of compressed charge.	Length of priming charge.	Length of sulphur plug.	Length of asphaltic composition, if any.	Length of sulphur filling.	Distance wires project below sulphur plug.	Distance from end of insulation to end of wires.
	In.	*In.*	*In.*	*In.*	*In.*	*In.*	*In.*	*In.*	*In.*	*In.*	*In.*
Western Coast	1.55	0.274	0.260	0.007	0.62	0.23	0.25		0.45	0.16	0.19
Special	1.75	.234	.220	.007	.56	.39	.25		.30	.16	.16
P. T. S. S.	1.55	.274	.260	.007	.28	.27	.25	0.50	.25	.12	.94
Foreign	1.55	.274	.260	.007	.44	.21	.25		.65	.16	.19

WEIGHT AND COMPOSITION OF CHARGES.

Following is a tabulation presenting the weight of the charges and their chemical composition as determined by analysis:

Weight and composition of charges of four No. 6 electric detonators.

Kind of electric detonator.	Weight of compressed charge.	Weight of priming charge.	Weight of total charge.	Constituents in compressed charge.				Constituents in priming charge.				Constituents in total charge.						
				Mercury fulminate.	Chlorate of potash.	Nitrocellulose.	Nitromannite.	Guncotton.	Mercury fulminate.	Picric acid.	Chlorate of potash.	Mercury fulminate.	Chlorate of potash.	Guncotton.	Nitrocellulose.	Nitromannite.	Picric acid.	Total.
	Grams.	*Grams.*	*Grams*	*Per ct.*	*Per ct.*	*Per ct.*	*Per ct.*	*Per ct.*	*Per ct.*	*Per ct.*	*Per ct.*	*Per ct.*	*Per ct.*	*Per ct.*	*Per ct.*	*Per ct.*	*Per ct.*	*Per ct.*
Western Coast a	0.8527	0.0155	0.8682	45.57	19.72	13.82	20.89	100.00	......	......	......	44.76	19.37	1.79	13.57	20.51	......	100.00
Special a	.9083	.0200	.9283	81.49	18.51	......	......	100.00	......	......	......	79.74	18.11	2.15	......	......	......	100.00
P. T. S. S. b	.6485	.3510	.9995	88.82	11.18	......	......	......	100.00	......	......	92.75	7.25	......	......	......	......	100.00
Foreign a	1.0108	.1640	1.1748	79.47	20.53	......	......	......	......	40.54	59.46	68.38	25.96	......	......	......	5.66	100.00

a J. H. Hunter, analyst.   b W. C. Cope, analyst.

RESULTS OF CALORIMETER TESTS.

The results of calorimeter tests of the four kinds of No. 6 electric detonator are tabulated below.

Results of calorimeter tests of four No. 6 electric detonators.

Kind of electric detonator.	Number of electric detonators used in each test.	Number of tests averaged.	Average heat evolved per electric detonator.	Total charge per electric detonator.	Average heat evolved per electric detonator had each contained the same weight of a charge consisting of 77.7 per cent of mercury fulminate and 22.3 per cent of chlorate of potash (exact combustion).[a]
			Large calories.	*Grams.*	*Large calories.*
Western Coast..................	15	2	[b] 0.95	0.8682	0.61
Special.......................	15	2	.75	.9283	.66
P. T. S. S.....................	15	2	.62	.9995	.71
Foreign.......................	15	3	[c] 1.12	1.1748	.83

[a] Berthelot, M., Explosives and their power, 1892, p. 470.
[b] This unusually high value is partly due to the high heat of total combustion of nitrocellulose (about three times that of mercury fulminate).
[c] This unusually high value is partly due to the high heat of total combustion of picric acid (about four times that of mercury fulminate).

SQUIRTED LEAD BLOCK TESTS.

The results of the squirted lead block tests are given herewith.

Results of squirted lead block tests[a] of four No. 6 electric detonators.

Kind of electric detonator.	Test No.	Volume of bore hole.		Increase of volume.	Average increase of volume.	Weight of total charge.
		Before test.	After test.			
		C. c.	*C. c.*	*C. c.*	*C. c.*	*Grams.*
Western Coast...............	AA14 / AA15	1.7 / 1.7	27.7 / 28.7	26.0 / 27.0	26.5	0.8682
Special......................	AA 9 / AA41	1.4 / 1.5	20.6 / 19.3	19.2 / 18.8	19.0	.9283
P. T. S. S....................	AA10 / AA11	1.7 / 1.7	20.0 / 19.8	18.3 / 18.1	18.2	.9995
Foreign......................	AA12 / AA26	1.7 / 1.7	28.9 / 28.6	27.2 / 26.9	27.0	1.1748

[a] For a description of the procedure in these tests, see p. 20.

CAST LEAD BLOCK TESTS.

Following are tabulated the results (Pl. VI) of cast lead block tests of the four kinds of No. 6 electric detonators:

Results of cast lead block tests of four No. 6 electric detonators.

Kind of electric detonator.	Test No.	Volume of bore hole.		Increase of volume.	Average increase of volume.	Weight of total charge.
		Before test.	After test.			
		C. c.	*C. c.*	*C. c.*	*C. c.*	*Grams.*
Western Coast................{	AA 6	1.8	22.7	20.9	21.2	0.8682
	AA31	1.7	23.2	21.5		
Special....................{	AA33	1.4	16.1	14.7	14.2	.9283
	AA58	1.4	15.2	13.8		
P. T. S. S................{	AA30	1.7	16.0	14.3	14.0	.9995
	AA55	1.6	15.2	13.6		
Foreign...................{	AA 5	1.7	19.7	18.0	18.2	1.1748
	AA54	1.7	20.0	18.3		

TESTS WITH LEAD PLATES.

Two series of tests of the four No. 6 electric detonators were made by the use of $\frac{1}{2}$-inch lead plates. In one series the electric detonators were placed on end on the plates and were detonated, the resultant depression of the plates being recorded. In the other series each electric detonator was placed on its side on the lead plate before detonation.

DETONATORS ON END.

The results of the lead-plate tests in which the detonators were placed on end (Pl. VII) are tabulated below:

Results of lead-plate tests of four No. 6 electric detonators, detonators being placed on end.

Kind of electric detonator.	Test No.	Volume of water contained in depression.	Diameter of crater.	Depth of crater.	Height of cone on bottom.
		C. c.	*Mm.*	*Mm.*	*Mm.*
Western Coast..........................	M155	0.15	11	5	Slight.
Special...............................	M159	.25	11	6	2
P. T. S. S.............................	M149	.40	13	7	3
Foreign...............................	M153	.45	13	7	3

Results of lead-plate tests of four grades of electric detonators, detonators being placed on side.

Kind of electric detonator.	Test No.	Volume of water contained in depression.	Diameter of crater.	Depth of crater.	Height of cone on bottom.
Western Coast..........................	M156	0.45	26	12	4
Special...............................	M160	.50	19	11	4
P. T. S. S.............................	a M150	.50	21	13	5
Foreign...............................	M154	.50	22	13	5

a Bottom of plate slightly raised; not raised in other tests.

RESULTS OF CAST LEAD BLOCK TESTS OF FOUR NO. 6 ELECTRIC DETONATORS. a, LEAD BLOCK BEFORE TEST; b, WESTERN COAST; c, SPECIAL; d, P. T. S. S.; e, FOREIGN.

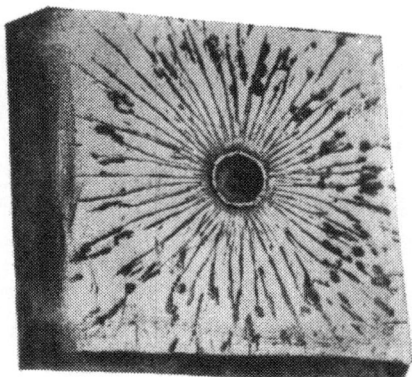

WESTERN COAST.

SPECIAL.

P. T. S. S.

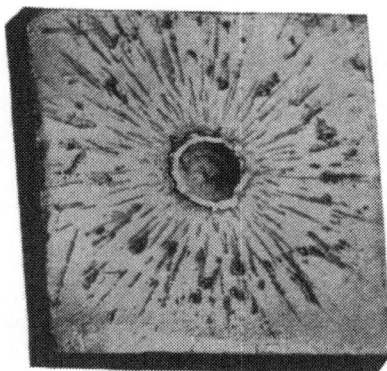

FOREIGN

SCORING OF LEAD PLATES BY FOUR NO. 6 ELECTRIC DETONATORS PLACED ON END.

DETONATORS ON SIDE.

Following are tabulated the results when the detonators were placed on their side (Pl. V, *C*) on the lead plates before detonation:

A second series of tests with the ½-inch lead plates, the electric detonators being fired on their side, was made, and the resultant depressions of the plates were measured with sand. The results are tabulated below:

Depression of ½-inch lead plates when electric detonators were fired on their side, depression measured with sand.

Kind of electric detonator.	Test No.	Plate No.	Weight of sand contained in depression, measurement No.—					Average.	Grand average.	Volume.a
			1	2	3	4	5			
			Grams.	*Grams.*	*Grams.*	*Grams.*	*Grams.*	*Grams.*	*Grams.*	*C. c.*
Western Coast....	M307	1	0.535	0.565	0.544	0.551	0.553	0.550	} 0.540	0.380
		2	.591	.601	.602	.560	.602	.591		
		3	.476	.470	.489	.472	.485	.478		
Special...........	M307	1	.507	.557	.562	.540	.587	.551	} .556	.392
		2	.551	.563	.566	.573	.590	.569		
		3	.540	.536	.568	.542	.551	.547		
P. T. S. S.........	M301	1	.574	.565	.620	.620	.590	.594	} .596	.420
		2	.580	.586	.607	.569	.596	.588		
		3	.622	.600	.615	.601	.598	.607		
Foreign...........	M307	1	.635	.668	.678	.684	.573	.648	} .595	.419
		2	.602	.584	.594	.587	.610	.595		
		3	.580	.560	.540	.511	.520	.542		

a The volume was computed from the grand average by dividing this by the specific gravity of the sand—in this case 1.42.

The results of the tests are fairly satisfactory, as they practically agree with the explosive efficiency established for electric detonators by the indirect methods.

NAIL TESTS.

The nail tests previously described were also used in connection with the investigation of the four grades of No. 6 electric detonators. The results (Pl. IV, *B*) are tabulated below:

Results of nail tests of four No. 6 electric detonators.

Kind of electric detonator.	Test No.	Angle of bending resulting from trial No.—					Average.	Minimum.
		1	2	3	4	5		
		°	°	°	°	°	°	°
Western Coast........................	M286	22	24	20	28	27	24.2	20
Special..............................	M287	16	35	17	16	23	21.4	16
P. T. S. S...........................	M288	23	24	25	24	26	24.4	23
Foreign.............................	M300	17	31	18	19	18	20.6	17

RATE-OF-DETONATION TESTS.

Rate-of-detonation tests similar to those with the different grades of P. T. S. S. electric detonators were conducted with the four No. 6 electric detonators. The results, according to the explosive used, are presented below.

TESTS WITH AN EXPLOSIVE OF CLASS 1, SUBCLASS *a*.

Following are the results of tests with an explosive of class 1, subclass *a* (an ammonium-nitrate explosive containing a sensitizer that is itself a sensitizer). The diameter of the cartridges used was seven-eighths of an inch.

Results of rate-of-detonation tests with an explosive of class 1, subclass a.

Kind of electric detonator.	Test No.	Rate of detonation in tube.							
		First quarter.		Second quarter.		Second half.		Full length.	
		Individual rate.	Average rate.	Individual rate.	Average rate.	Individual rate.	Average rate.	Individual rate.	Average rate.
		Meters per second.	*Meters per second.*	*Meters per second.*	*Meters per second.*	*Meters per second.*	*Meters per second.*	*Meters per second.*	*Meters per second.*
Western Coast...............	D970 D971	2,343 2,295	} 2,319	{ 2,295 { 2,250	} 2,272	{ 2,585 { 2,556	} 2,570	{ 2,445 { 2,406	} 2,426
Special....................	D976 D978 D979	1,891 2,419 2,393	} 2,234	{ 2,296 { 2,393 { 2,206	} 2,298	{ 2,761 { 2,459 { 2,419	} 2,546	{ 2,368 { 2,432 { 2,356	} 2,385
P. T. S. S...................	D966 D967	2,445 2,472	} 2,458	{ 2,678 { 2,586	} 2,632	{ 2,205 { 2,205	} 2,205	{ 2,368 { 2,356	} 2,362
Foreign....................	D974 D973	2,778 2,143	} 2,460	{ 1,814 { 2,393	} 2,104	{ 2,795 { 2,866	} 2,830	{ 2,459 { 2,528	} 2,494
Grand average.........	....	....	2,393	....	2,326	....	2,538	....	2,417

The average rate for the meter length was practically uniform, but such difference as was shown indicated that the ascending order of explosive efficiency of the detonators is as follows: P. T. S. S., special, Western Coast, foreign.

The percentage of complete detonations with each detonator was as follows:

Percentage of complete detonations with an explosive of class 1, subclass a.

Kind of electric detonator.	Number of tests.	Number of tests in which incomplete detonation occurred.	Percentage of complete detonations.
			Per cent.
Western Coast...................................	2	0	100
Special..	5	0	100
P. T. S. S......................................	2	0	100
Foreign.......................................	3	1	67

TESTS WITH AN EXPLOSIVE OF CLASS 1, SUBCLASS *b*.

Two rate-of-detonation tests were made of each of the four kinds of No. 6 electric detonators on an explosive of class 1, subclass *b* (an ammonium-nitrate explosive containing a sensitizer that is not itself an explosive) being used. The cartridges used were 1¾ inches in diameter. In no test did detonation occur, so that the tests failed to discriminate between the different kinds of electric detonators.

TESTS WITH A 20 PER CENT "STRAIGHT" NITROGLYCERIN DYNAMITE.

The results of tests with a 20 per cent "straight" nitroglycerin dynamite are presented below. The diameter of the cartridges was seven-eighths of an inch.

Results of rate-of-detonation tests with a 20 per cent "straight" nitroglycerin dynamite.

Kind of electric detonator.	Test No.	Rate of detonation in tube.							
		First quarter.		Second quarter.		Second half.		Full length.	
		Individual rate.	Average rate.	Individual rate.	Average rate.	Individual rate.	Average rate.	Individual rate.	Average rate.
		Meters per second.	*Meters per second.*	*Meters per second.*	*Meters per second.*	*Meters per second.*	*Meters per second.*	*Meters per second.*	*Meters per second.*
Western Coast..............	D996 / D997	2,778 / 3,462	} 3,120	{ 2,960 / 2,679	} 2,820	{ 3,147 / 3,285	} 3,216	{ 3,000 / 3,147	} 3,074
Special....................	D998 / D999	3,048 / 3,299	} 3,174	{ 2,928 / 2,587	} 2,758	{ 3,069 / 3,027	} 3,048	{ 3,027 / 2,967	} 2,997
P. T. S. S..................	D1000 / D1002	3,729 / 3,947	} 3,838	{ 2,683 / 2,443	} 2,563	{ 2,767 / 3,309	} 3,038	{ 2,933 / 3,158	} 3,046
Foreign....................	D994 / D995	2,922 / 3,000	} 2,961	{ 3,125 / 2,557	} 2,841	{ 2,866 / 3,061	} 2,964	{ 2,941 / 2,903	} 2,922
Grand average........	....	....	3,273	....	2,746	....	3,066	....	3,010

The figures representing the grand averages indicate that the rate was influenced by a negative acceleration in the second quarter meter followed by a positive acceleration in the second half meter. This acceleration occurred in all tests except D996 and D994.

The uniformity of the rates for the last half meter and the meter is noteworthy.

With a 20 per cent "straight" nitroglycerin dynamite such difference as was shown in the tests indicated that the ascending order of explosive efficiency is: Foreign, special, P. T. S. S., Western Coast.

TESTS WITH A 40 PER CENT STRENGTH AMMONIA DYNAMITE CONTAINING NITROSUBSTITUTION COMPOUNDS.

Following are tabulated the results of tests with a 40 per cent ammonia dynamite containing nitrosubstitution compounds. The explosive was repacked in cartridges seven-eighths of an inch in diameter.

Results of rate-of-detonation tests with a 40 per cent strength ammonia dynamite containing nitrosubstitution compounds.

Kind of electric detonator.	Test No.	Rate of detonation in tube.							
		First quarter.		Second quarter.		Second half.		Full length.	
		Individual rate.	Average rate.	Individual rate.	Average rate.	Individual rate.	Average rate.	Individual rate.	Average rate.
		Meters per second.	*Meters per second.*	*Meters per second.*	*Meters per second.*	*Meters per second.*	*Meters per second.*	*Meters per second.*	*Meters per second.*
Western coast.............	D913 D923	2,557 3,082	2,820	2,394 2,500	2,447	2,812 2,616	2,714	2,632 2,687	2,660
Special..................	D879 D914	2,820 2,587	2,704	2,588 2,781	2,684	2,900 3,048	2,974	2,794 2,853	2,824
P. T. S. S...............	D904 D919 D920	2,368 2,472 2,250	2,363	2,296 3,041 2,446	2,594	2,572 2,663 2,543	2,593	2,446 2,695 2,439	2,527
Foreign.................	D912 D922	2,250 2,616	2,433	3,129 3,125	3,127	2,446 2,557	2,502	2,528 2,542	2,535

The rate for the second quarter meter was the highest for the P. T. S. S. and the foreign electric detonators.

With a 40 per cent strength ammonia dynamite containing nitrosubstitution compounds the tests indicated that the ascending order of explosive efficiency is: P. T. S. S., foreign, Western Coast, special.

TESTS WITH A 35 PER CENT STRENGTH GELATIN DYNAMITE 2 YEARS OLD.

The results of tests with a 35 per cent strength gelatin dynamite (two years old) are tabulated below. The diameter of the cartridges used was 1¼ inches.

Results of rate-of-detonation tests with a 35 per cent strength gelatin dynamite two years old.

Kind of electric detonator.	Test No.	Length a of part of file blown off or detonated.	Percentage of file that detonated.	Average.	Remarks.
		Inches.	*Per cent.*	*Per cent.*	
Western Coast.......................	D897 D898	7.0 8.0	17 19	18.0	Partial detonation. Do.
Special...........................	D901 D902	10.0 9.0	24 21	22.5	Do. Do.
P. T. S. S.......................	D889 D896 D958	7.0 18.0 17.0	0 43 40	27.5	No detonation. Partial detonation. Do.
Foreign.........................	D899 D900	6.0 7.0	14 17	15.5	Do. Do.

a Full length of file, 42 inches.

The evidence of no detonation in test D889 was that nothing but the noise of the electric detonator was audible when the trial was made.

With the two-year-old sample of 35 per cent strength gelatin dynamite used the tests indicated that the ascending order of explosive efficiency is: Foreign, Western Coast, special, P. T. S. S.

The percentage of partial detonations with each electric detonator was as follows:

Percentage of partial detonations with a 35 per cent strength gelatin dynamite two years old.

Kind of electric detonator.	Number of tests.	Number of tests in which partial detonation occurred.	Percentage of complete detonations.
			Per cent.
Western Coast...	2	2	100
Special...	2	2	100
P. T. S. S...	3	2	67
Foreign..	2	2	100

TESTS WITH A 40 PER CENT STRENGTH GELATIN DYNAMITE, FROZEN.

Following are the results of tests with a 40 per cent strength gelatin dynamite (frozen). The diameter of the cartridges used was 1¼ inches.

Results of rate-of-detonation tests with a 40 per cent strength gelatin dynamite, frozen.

Kind of electric detonator.	Test No.	Rate of detonation in tube.			
		First quarter.	Second quarter.	Second half.	Full length.
		Meters per second.	*Meters per second.*	*Meters per second.*	*Meters per second.*
Western Coast.............................	D1091	3,273	7,177	5,705	5,028
Special.....................................	D1092	4,167	6,357	6,013	5,460
P. T. S. S..................................	D1093	4,167	7,759	6,522	5,921
Foreign....................................	D1090	3,090	14,833	5,361	5,235
Grand average...........................	....	3,674	9,032	5,900	5,411

The figures included in the "grand average" show that the maximum rate occurred in the second quarter, with a subsequent falling off in the rate; moreover, the rate varied similarly in each individual test.

With the explosive used in the tests the results indicate that the ascending order of explosive efficiency is: Western Coast, foreign, special, P. T. S. S.

TESTS WITH A 35 PER CENT STRENGTH GELATIN DYNAMITE 3 YEARS OLD.

One test each of the Western Coast, the special, and the foreign No. 6 electric detonators was made with a 35 per cent strength gelatine dynamite 3 years old. The diameter of the cartridges used was 1½ inches. No detonation took place in any of the tests, as the explosive was so old and insensitive to detonation that for the purpose of discriminating between detonators it was useless.

SMALL LEAD BLOCK TESTS.

Small lead block tests were made with the four No. 6 electric detonators. The results, according to the explosive tested, are given below.

TESTS WITH A 20 PER CENT "STRAIGHT" NITROGLYCERIN DYNAMITE.

Three series of tests were conducted with a 20 per cent "straight" nitroglycerin dynamite in different conditions as indicated below.

Results of small lead block tests with a 20 per cent "straight" nitroglycerin dynamite with 6 per cent of added water.

Kind of electric detonator.	Test No.	Compression.	Average compression.
		Mm.	*Mm.*
Western Coast	B759 B768 B777	15.00 14.75 15.25	15.00
Special	B760 B769 B778	14.00 15.25 15.25	14.83
P. T. S. S.	B761 B770 B779	15.00 15.00 15.00	15.00
Foreign	B758 B767 B776	14.25 14.75 15.00	14.67

In the tests the explosive produced nearly uniform individual compressions and little difference in the average compressions, but such difference as was shown indicated that the ascending order of explosive efficiency is: Foreign, special, Western Coast, P. T. S. S.

The results of tests with the same explosive, but containing 4 per cent of added water and frozen, were as follows:

Results of small lead-block tests with a 20 per cent "straight" nitroglycerin dynamite containing 4 per cent of added water and frozen.

Kind of electric detonator.	Test No.	Temperature of frozen explosive.	Percentage of water added.	Compression.	Average compression.
		° *C.*		*Mm.*	*Mm.*
Western Coast	B651	−9.0	4.0	12.75	12.75
Special	B652	−9.0	4.0	12.50	12.50
P. T. S. S.	B653	−9.0	4.0	13.00	13.00
Foreign	B650	−9.0	4.0	12.75	12.75

With the explosive in the condition mentioned, the results of the tests indicate that the ascending order of explosive efficiency is: Special, Western Coast, foreign, and P. T. S. S.

Further tests were conducted with 6 per cent of water added to the explosive and the explosive frozen (temperature 9° C.). Three tests were made with each of the four grades of electric detonators, but the explosive was too insensitive to detonation to be discriminative, as no compression of any of the blocks was produced.

TESTS WITH A 40 PER CENT STRENGTH AMMONIA DYNAMITE.

Following are the results of tests with a 40 per cent ammonia dynamite, to which had been added 6 per cent of water:

Results of small lead block tests with a 40 per cent strength ammonia dynamite containing 6 per cent of added water.

Kind of electric detonator.	Test No.	Compression.	Average compression.
		Mm.	*Mm.*
Western Coast	B660	8.25	
	B669	9.00	
	B678	9.00	8.75
	B687	8.00	
	B696	9.50	
Special	B661	8.00	
	B670	8.25	
	B679	8.50	8.40
	B688	7.75	
	B697	9.25	
P. T. S. S.	B662	9.75	
	B671	8.75	
	B680	9.25	8.95
	B689	7.75	
	B698	9.25	
Foreign	B659	8.00	
	B668	10.00	
	B677	9.25	9.20
	B686	9.50	
	B695	9.25	

The results were obviously erratic. However, the tests indicated that the ascending order of explosive efficiency is: Special, Western Coast, P. T. S. S., foreign.

TESTS WITH A 40 PER CENT STRENGTH GELATIN DYNAMITE, FROZEN.

The results of tests with a 40 per cent strength gelatin dynamite (frozen) were as follows (Pl. V, *B*):

Results of small lead block tests with a 40 per cent strength gelatin dynamite, frozen.

Kind of electric detonator.	Test No.	Temperature of frozen explosive.	Compression.	Average compression.
		° C.	*Mm.*	*Mm.*
Western Coast	B635	−2.5	14.75	
	B642	−5.0	17.75	
	B705	+ .5	12.50	13.90
	B714	+2.5	12.50	
	B723	+2.5	12.00	
Special	B636	−2.5	12.75	
	B643	−5.0	18.00	
	B706	+ .5	15.00	14.85
	B715	+2.5	14.75	
	B724	+2.5	13.75	
P. T. S. S.	B631	−4.5	12.50	
	B644	−5.0	19.25	
	B707	+ .5	14.25	14.95
	B716	+2.5	14.25	
	B725	+2.5	14.50	
Foreign	B634	−2.5	11.00	
	B641	−5.0	18.00	
	B704	+ .5	15.00	15.15
	B713	+2.5	16.25	
	B722	+2.5	15.50	

As indicated by the table, the results of the tests were very erratic with this frozen gelatin dynamite. The insensitiveness of this explosive has been mentioned in a foregoing section regarding the incompleteness of detonation in tests with the Nos. 3, 4, and 5 electric detonators. With the No. 6 electric detonators, however, detonation was complete in every trial.

EXPLOSION-BY-INFLUENCE TESTS.

Tests involving explosion by influence as outlined in a foregoing section relative to tests of different grades of P. T. S. S. electric detonators were made of the four kinds of No. 6 electric detonators, as described below:

TESTS WITH AN EXPLOSIVE OF CLASS 1, SUBCLASS a.

Following are the results of tests with an explosive of class 1, subclass a (an ammonium-nitrate explosive containing a sensitizer that is itself an explosive). The size of the cartridges used was $1\frac{1}{4}$ by 8 inches and the average weight was 166 grams.

Results of explosion-by-influence tests with an explosive of class 1, subclass a (sample 1).

Kind of electric detonator.	Test No.	Distance separating cartridges.	Result on upper cartridge.	Established distance at which detonation did not occur.
		Inches.		*Inches.*
Western Coast.......................	J758	2	Did not explode.......	
	J759	1	Exploded.............	2
	J760	2	Did not explode......	
Special........................	J761	1	Exploded.............	
	J762	2	Did not explode......	2
	J763	2	.....do...............	
P. T. S. S........................	J741	4	.....do...............	
	J742	3	.....do...............	
	J743	2	.....do...............	
	J744	1	Exploded.............	3
	J745	2	.....do...............	
	J746	3	Did not explode......	
Foreign........................	J753	3	.....do...............	
	J754	2	.....do...............	
	J755	1	.....do...............	1
	J756	0	Exploded.............	
	J757	1	Did not explode......	

With the ammonium-nitrate explosive used, the results of the tests indicated that the ascending order of explosive efficiency is: Foreign, Western Coast and special, P. T. S. S.

TESTS WITH AN EXPLOSIVE OF CLASS 4.

The results of tests with an explosive of class 4 (an explosive in which the characteristic material is nitroglycerin) are tabulated below. The size of the cartridges used was $1\frac{1}{4}$ by 8 inches, their average weight being 161 grams, except that in the trials under test J896 the upper cartridge weighed 110 grams and was 5 inches long.

Results of explosion-by-influence tests with an explosive of class 4.

Kind of electric detonator.	Test No.	Distance separating cartridges.	Result on upper cartridge.
		Inches.	
Western Coast................................	J895	5	Exploded.
	J895	5	Did not explode.
	J895	5	Do.
	J895	5	Do.
	J896	4	Exploded.
	J896	4	Do.
	J896	4	Did not explode.
Special....................................	J895	5	Exploded.
	J895	5	Did not explode.
	J895	5	Do.
	J895	5	Do.
	J896	4	Exploded.
	J896	4	Do.
	J896	4	Did not explode.
P. T. S. S..............................	J895	5	Do.
	J895	5	Do.
	J895	5	Do.
	J895	5	Exploded.
	J896	4	Do.
	J896	4	Do.
	J896	4	Did not explode.
Foreign.................................	J895	5	Do.
	J895	5	Do.
	J895	5	Do.
	J895	5	Do.
	J896	4	Do.
	J896	4	Do.
	J896	4	Do.

Percentage of explosions of the upper cartridge in explosion-by-influence tests with an explosive of class 4.

Kind of electric detonator.	Number of tests.	Number of explosions of upper cartridge.	Percentage of explosions of upper cartridge.
			Per cent.
Western Coast................................	7	3	43
Special....................................	7	3	43
P. T. S. S................................	7	3	43
Foreign.................................	7	0	0

In all tests in which the distance separating cartridges was 5 inches the bottoms of the cartridges (as packed) faced each other; in all tests in which the distance was 4 inches the tops of the cartridges faced each other.

The tests indicated that the foreign electric detonator was not as effective under the conditions of the tests as were the other three.

TESTS WITH A 40 PER CENT STRENGTH AMMONIA DYNAMITE CONTAINING NITROSUBSTITUTION COMPOUNDS.

Following are the results of tests with a 40 per cent strength ammonia dynamite containing nitrosubstitution compounds. The cartridges used measured 1¼ by 8 inches and their average weight was 226 grams.

Results of explosion-by-influence tests with a 40 per cent strength ammonia dynamite containing nitrosubstitution compounds.

Kind of electric detonator.	Test No.	Distance separating cartridges.	Result on upper cartridge.	Distance at which detonation did not occur.
		Inches.		*Inches.*
Western Coast..........................	J714	8	Exploded.............	
	J715	9	Did not explode.......	9
	J716	9	.....do................	
Special................................	J717	8	.....do................	
	J718	7	Exploded.............	8
	J719	8	Did not explode.......	
P. T. S. S.............................	J689	14	.....do................	
	J690	12	.....do................	
	J691	9	.....do................	
	J692	7	.....do................	
	J693	6	.....do................	
	J694	4	Exploded.............	8
	J695	5	.....do................	
	J696	6	.....do................	
	J697	7	.....do................	
	J698	8	Did not explode.......	
	J699	8	.....do................	
Foreign................................	J711	8	.....do................	
	J712	7	Exploded.............	8
	J713	8	Did not explode.......	

The tests show practically the same result regardless of the electric detonator used.

TESTS WITH A 35 PER CENT STRENGTH GELATIN DYNAMITE 2 YEARS OLD.

The results of tests with a 35 per cent strength gelatin dynamite (two years old) are tabulated below. The cartridges used were $1\frac{1}{4}$ by 8 inches, their average weight being 265 grams.

Results of explosion-by-influence tests with a 35 per cent strength gelatin dynamite (two years old).

Kind of electric detonator.	Test No.	Distance separating cartridges.	Result on upper cartridge.
		Inches.	
Western Coast..	J734	0	Did not explode.
	J735	0	Do.
Special..	J737	0	Do.
	J738	0	Do.
P. T. S. S...	J724	6	Do.
	J725	5	Do.
	J726	4	Do.
	J727	2	Do.
	J728	0	Do.
	J729	0	Do.
Foreign...	J736	0	Do.
	J739	0	Do.

These tests failed to discriminate between the different detonators, as in no trial did the explosion of the lower cartridge cause the detonation of the upper cartridge.

TRAUZL LEAD BLOCK TESTS.[a]

In testing the four kinds of No. 6 electric detonators the Trauzl lead block tests were used in addition to the tests previously described.

The Trauzl lead blocks are cylindrical in shape, measuring 200 mm. in diameter and 200 mm. in height. They have an axial bore hole 25 mm. in diameter and 125 mm. in depth. The charge of 20 grams of the explosive in which the electric detonator was embedded was placed in the bottom of the bore hole and no stemming was used. The increase in the volume of water that the bore hole would contain after an explosion was the result recorded.

Following is a tabulation of results of Trauzl lead block tests in which a 20 per cent "straight" nitroglycerin dynamite was used. The charge of explosive in each test was 20 grams, to which was added 6 per cent of water.

Results of Trauzl lead block tests with a 20 per cent "straight" nitroglycerin dynamite.

Kind of electric detonator.	Test No.	Expansion.	Average expansion.
		C. c.	*C. c.*
Western Coast	A817 A818	175 173	174
Special	A819 A821	177 175	176
P. T. S. S.	A822 A824	178 175	176
Foreign	A815 A816	178 179	178

As indicated by the table, the average expansion of the blocks in each test was nearly the same.

PERCENTAGES OF DETONATIONS IN INDIRECT TESTS OF FOUR KINDS OF NO. 6 ELECTRIC DETONATORS.

The percentages of detonations in the indirect tests of the four kinds of No. 6 electric detonators are given below. The percentages of detonations in the tests of each electric detonator are also averaged, each average percentage having a value proportional to the number of tests from which it is computed; that is, each percentage is multiplied by the number of tests it represents and the sum of the products is divided by the total number of tests of the electric detonator considered.

---

a For a more extended description of this test see Bull. 15, Bureau of Mines: Investigations of explosives used in coal mines; with a chapter on the natural gas used at Pittsburgh, by G. A. Burrell, and an introduction by C. E. Munroe, by Clarence Hall, W. O. Snelling, and S. P. Howell, 1912, pp. 114-116.

Percentages of detonations in indirect tests of four kinds of No. 6 electric detonators.

Class and grade of explosive.	Character of test.	Western Coast.		Special.		P. T. S. S.		Foreign.	
		Percentage of detonations.	Number of tests.	Percentage of detonations.	Number of tests.	Percentage of detonations.	Number of tests.	Percentage of detonations.	Number of tests.
Class 1, subclass *a* (sample 1).	Rate of detonation.	*Per ct.* 100	2	*Per ct.* 100	5	*Per ct.* 100	2	*Per ct.* 67	3
40 per cent strength gelatin dynamite (frozen).	.....do.......	100	1	100	1	100	1	100	1
35 per cent strength gelatin dynamite (2 years old).	.....do.......	100	2	100	2	67	3	100	2
20 per cent "straight" nitroglycerin dynamite (frozen and containing 6 per cent of added water).	Small lead block.	0	3	0	3	0	3	0	3
40 per cent strength gelatin dynamite (frozen).	.....do.......	100	5	100	5	100	5	100	5
Average.........	....	55.5	....	71.4	....	63.2	....	61.1	....
Total number of tests..	....	....	18	....	21	....	19	....	18

COMPARATIVE EXPLOSIVE EFFICIENCY.

The percentage of explosive efficiency of the four kinds of No. 6 electric detonators was obtained by averaging all tests in which the rate of detonation, compression, or expansion was determined for all detonators. Each percentage was given a value proportional to the number of tests from which the percentage was computed. In each case the percentage of explosive efficiency of the P. T. S. S. No. 6 electric detonator is given a value of 100 and is taken as the unit of comparison.

Explosive efficiency of four No. 6 electric detonators of different makes.

Class of explosive.	Kind of test.	Western Coast.					Special.					P. T. S. S.					Foreign.				
		Rate. (Meters per sec.)	Compression. (Mm.)	Expansion. (Cu. mm.)	Percentage of explosive efficiency. (Per cent.)	Number of tests.	Rate. (Meters per sec.)	Compression. (Mm.)	Expansion. (Cu. mm.)	Percentage of explosive efficiency. (Per cent.)	Number of tests.	Rate. (Meters per sec.)	Compression. (Mm.)	Expansion. (Cu. mm.)	Percentage of explosive efficiency. (Per cent.)	Number of tests.	Rate. (Meters per sec.)	Compression. (Mm.)	Expansion. (Cu. mm.)	Percentage of explosive efficiency. (Per cent.)	Number of tests.
20 per cent "straight" nitroglycerin dynamite in $\frac{7}{8}$-inch cartridges.	Rate-of-detonation.	3,074			100.9	2	2,997			98.4	2	3,046			100.0	2	2,922			95.9	2
40 per cent strength ammonia dynamite (containing nitrosubstitution compounds) in $\frac{7}{8}$-inch cartridges.	do	2,660			105.3	2	2,824			111.8	2	2,527			100.0	3	2,535			100.3	2
Class 1, subclass a (sample 1) in $\frac{7}{8}$-inch cartridges.	do	2,426			102.7	2	2,385			101.0	3	2,362			100.0	2	2,494			105.5	2
40 per cent strength gelatin dynamite (frozen in 1¼-inch cartridges).	do	5,028			84.9	1	5,460			92.2	1	5,921			100.0	1	5,235			88.4	1
35 per cent strength gelatin dynamite (2 years old).	do	a 7.5	i5.00		42.9	2	a 9.5	14.83		54.3	2	a 17.5	15.00		100.0	2	a 6.5	14.67		37.1	2
20 per cent "straight" nitroglycerin dynamite (containing 6 per cent of added water).	Small lead block.				100.0	3				96.9	3				100.0	3				97.8	3
20 per cent "straight" nitroglycerin dynamite (containing 4 per cent of added water).	do		12.75		98.1	1		12.50		96.2	1		13.00		100.0	1		12.75		98.1	1
40 per cent strength gelatin dynamite (frozen).	do		13.90		93.0	5		14.85		99.3	5		14.95		100.0	5		15.15		101.3	5
40 per cent strength ammonia dynamite (containing 6 per cent of added water).	do		8.75		97.8	5		8.40		93.9	5		8.95		100.0	5		9.20		102.8	5
20 per cent "straight" nitroglycerin dynamite (containing 6 per cent of added water).	Trauzl lead block.			174	98.9	2			176	100.0	2			176	100.0	2			178	101.1	2
Average.					93.5					95.5					100.0					95.2	
Total number of tests.						25					26					20					25

a Inches that detonated.

COMPARATIVE EXPLOSIVE EFFICIENCY OF FOUR KINDS OF NO. 6 ELECTRIC DETONATORS.

The comparative explosive efficiency of the four grades of electric detonators (fig. 5), as established, is tabulated below.

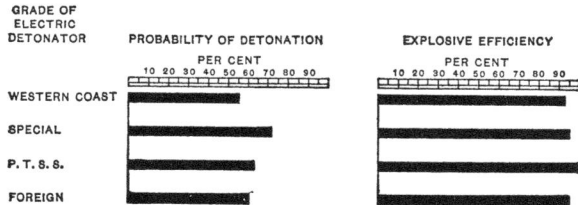

FIGURE 5.—Comparative explosive efficiency of four kinds of No. 6 electric detonators as established by indirect tests.

Explosive efficiency of four kinds of No. 6 electric detonators.

Kind of electric detonator.	Percentage of probability of detonation.	Explosive efficiency for those tests in which detonation occurred.
	Per cent.	*Per cent.*
Western Coast	55.5	93.5
Special	71.4	95.5
P. T. S. S. No. 6	63.2	100.0
Foreign	61.1	95.2

RELATIVE STRENGTH OF DETONATORS AND ELECTRIC DETONATORS.

It is generally recognized that the safest way of firing shots in blasting operations is with electric detonators by means of the electric current. This is especially true in gaseous coal mines, because if fuse is used the flame produced by the burning fuse may ignite such inflammable gaseous mixtures as are present. There is also danger of a hangfire when the charge may be exploded unexpectedly, due to the smoldering of the fuse.

There are, however, many conditions of mining under which electric detonators can not be used advantageously. In driving drifts in many of the metal mines of this country fuse is generally used. In work of this kind it is often necessary to fire dependent shots, and the flying rock from one shot may disconnect or cause short-circuiting of the electric wires of the detonators wired for succeeding shots. When fuse is used, different lengths can be cut and, before lighting, the projecting ends can be coiled in a place within the mouth of the hole where they are well protected.

It has been observed that mercury fulminate ignited in small quantities develops its full force only when confined. It also has been believed that the sulphur plug in an electric detonator offers more confinement to the fulminating composition of a detonator than a piece of fuse does, even when the fuse is properly used and securely crimped in place. Therefore it seemed desirable to make comparative efficiency tests of both electric detonators and detonators fitted with fuse. The nail test was adopted for the reason that it produced more nearly the results established for the efficiency of detonators than any of the other direct methods.

The nail test was made with the No. 3 and the No. 6 detonators and with the electric detonators made from these detonators, with the following results:

Results of nail tests with No. 3 and No. 6 detonators and electric detonators.

Grade of detonator or electric detonator.	Test No.	Degrees of bending in trial—					Aver-age.	Mini-mum.
		1	2	3	4	5		
		°	°	°	°	°	°	°
No. 3 detonator *a*.....................................	M289	8	7	9	9	8	8.2	7
No. 3 electric detonator................................	M279	12	10	8	9	7	9.2	7
No. 6 detonator *a*.....................................	M318	32	29	35	30	31	31.4	29
No. 6 electric detonator................................	M321	31	31	33	36	27	31.6	27

a Fired with fuse placed against the compressed charge of mercury fulminate composition and crimped in place.

The compressed charge of the No. 6 electric detonator weighed 1.0225 grams and consisted of 89.61 per cent of mercury fulminate and 10.39 per cent of potassium chlorate. The priming charge consisted of 0.02 gram of loose guncotton. As the weight of charge of this electric detonator was practically the same as of the P. T. S. S. electric detonator, the increased strength as shown by the nail test (nail bent 31.6° by the No. 6 electric detonator as compared with 24.4° by the P. T. S. S. electric detonator) would indicate that the quantity of compressed charge in the detonator may be a function of the efficiency of the detonator.

The tests showed that with low-grade detonators the strength of electric detonators is slightly greater than that of the corresponding detonators, but that with greater charges, such as the No. 6 detonators contain, the strength of the two types is practically the same. This indicates that the additional confinement given by the plug of the electric detonator as compared with the fuse of the fuse detonator is important only with the low-grade detonators.

A serious objection to the use of fuse and detonators in wet blasting is the fact that it is impossible to perfect a waterproof seal at the top of the detonator when it is crimped on the fuse. The ordinary fuse crimper depends on flattening the sides of the copper shell to contract the diameter of the detonator. Tests have shown that when a detonator is crimped on a fuse in this manner and submerged under water for 30 minutes the fulminating charge and the powder train at the end of the fuse in the detonator become damp. The spit of the lighted fuse, if the fuse burns through, is usually of insufficient intensity to cause an explosion of the fulminating charge. In some cases a sharp explosion or an explosion of a very low order occurs. If only a little water enters the detonator, the spit of the burning fuse is often sufficient to cause the fulminating charge to detonate with a sharp report and completely destroy the copper shell. In some of the tests 70 to 80 per cent of the compressed fulminating charge was recovered in the lower part of the copper shell. The spit of the burning fuse had seemingly caused a part of the fulminating composition to detonate, the detonation destroying the top part of the copper shell but not being propagated throughout the remainder of the wet fulminating charge. In these instances a slight report only was audible. Obviously an explosion of this order would not cause a complete detonation of dynamite or other high explosives. In some of the tests a thin coating of tallow was placed on the fuse one-fourth of an inch from the end and extending a distance of one-half of an inch up on the fuse. In the tests in which tallow was placed around the fuse before it was inserted into the detonator a more perfect seal was made.

A new crimper recently placed on the market crimps the detonator on the fuse in a manner different from that of any of the other types of crimpers. The salient feature of this crimper is its ability to contract the top of the detonator uniformly and to form a $\frac{1}{8}$-inch groove around the copper shell, thus perfecting a seal of the detonator on the fuse that will permit submersion under water for 30 minutes. The shell is pressed firmly and uniformly into the fuse, but not far enough to break or separate the powder train. Owing to the varying diameters of different types of fuse and the probability of considerable variation in the same type or even in the same coil of fuse, the use of a thin film of tallow around the end of the fuse that is inserted into the detonator, as described above, will make a better seal irrespective of the crimper used.

TESTS WITH A TRINITROTOLUENE DETONATING FUSE.

As the results of all tests made with explosives sensitive to detonation showed that when a complete detonation was obtained the rate of detonation was practically the same, the authors decided to carry on tests with a few explosives, using No. 6 electric detonators and trinitrotoluene detonating fuse as the initiatory explosive.

The trinitrotoluene detonating fuse used in the tests was a lead tube filled with trinitrotoluene, and is commercially known as "cordeau detonant." The results of physical examination of the fuse were as follows.

Results of physical examination of 6-mm. detonating fuse (cordeau detonant).

Outside diameter, inches... 0. 2275
Thickness of lead, inches.. . 0275
Inside diameter of tube, inches.................................. . 1725
Weight of a foot length, grams.................................. 41. 74
Weight of a foot length of lead tube, grams..................... 35. 32
Weight of a foot length of charge, grams....................... 6. 42
Density of charge... 1. 40
Consistency of charge: Powdered; very fine; dry; soft; slightly cohesive.
Color of charge: Straw.

The tests were made with explosives in which a 6-inch length of the detonating fuse (cordeau detonant) was embedded centrally at one end of the charge, the side of the fuse being slit and spread open from one end a distance of 1¼ inches. A No. 6 electric detonator was placed against the trinitrotoluene in the slit and tied firmly in place. The electric detonator and attached detonating fuse (cordeau detonant) was imbedded in the explosive. Following is a tabulation of the results of the tests:

Results of rate-of-detonation tests of explosives with a No. 6 electric detonator and detonating fuse (cordeau detonant).

Class of explosive.	Test No.	Rate of detonation.	Average rate of detonation.
		Meters per second.	*Meters per second.*
Class 1, subclass *a* (sample 1)...................	D981 D982	2,231 2,225	} 2,228
Class 1, subclass *b* (sample 2)...................	D990 D991	(*a*) (*a*)	}..........
20 per cent "straight" nitroglycerin dynamite...................	D1007 D1008 D1009	3,156 2,947 3,190	} 3,098
40 per cent strength ammonia dynamite (containing nitrosubstitution compounds)...................	D880 D883 D884 D885 D886	2,444 (*a*) 2,945 2,821 2,713	} 2,731
35 per cent strength gelatin dynamite (two years old)...............	D893 D894	(*a*) (*a*)	}..........

a Incomplete detonation; rate not determined.

Comparative results of tests with detonating fuse fired with No. 6 detonators and with No. 6 electric detonators used alone.

	No. 6 electric detonator and detonating fuse.	No. 6 electric detonator.
Averages of the rate of detonation of three explosives, meters per second.........	2,686	2,645
Explosive efficiency, per cent.......................	101. 6	100. 0

The results of the tests show that a 6-inch length of detonating fuse (cordeau detonant) used in connection with a No. 6 electric detonator does not increase the rate of detonation of the explosives tested.   The slight increase indicated in the table is explained by the fact that the rate of detonation of detonating fuse (cordeau detonant) itself is about 4,900 meters per second and that the fuse extended about one-eighth the length of the charge.

If the detonating fuse had extended the full length of the charge the rate of detonation of the explosive would probably have been increased to 4,900 meters per second, the rate of the detonating fuse.

Detonating fuse has been used to some extent in deep-hole blasting. A piece of the fuse is laid beside the charge of high explosive that has been inserted into the hole.   The fuse, when detonated, accelerates the rate of detonation of the explosive, thus producing a greater shattering effect on the surrounding rock.   Detonating fuse has also been used to replace electric detonators in large blasts when simultaneous blasting is desired.   Obviously, when detonating fuse is used in drill holes containing a long charge of explosive whose rate of detonation is less than that of the detonating fuse, a greater shattering effect will be produced.   When the rate of detonation of the explosive charge is greater than that of the detonating fuse, the only advantage in using the fuse would be to insure a complete detonation of the entire charge of explosive.

TESTS WITH DETONATORS DISTRIBUTED IN CHARGE.

With long charges of high explosives in blasting work, it has sometimes been the custom to place detonators at intervals in the charges, in the belief that the work accomplished by the explosives would be increased.   Rate-of-detonation tests were made with an explosive of class 1, subclass *a*, sample 3 (an ammonium-nitrate explosive containing a sensitizer that is itself an explosive), in charges $1\frac{1}{4}$ inches in diameter, with and without No. 7 detonators distributed in the explosive, to determine whether detonation of the charge would occur at a greater distance because of the presence of the detonators.   The explosive had been previously tested (see p. 29) in charges $1\frac{1}{2}$ inches in diameter, with the result that the No. 7 electric detonator caused complete detonation in every trial, whereas the No. 3 electric detonator failed to do so once out of three trials.   The explosive was insensitive to detonation and was purposely chosen for this reason. The results were as follows:

Results of rate-of-detonation tests in which No. 7 detonators were distributed in the charge.

Grade of detonator.	Test No.	Dimensions of galvanized-iron tube used.		Result.
		Diameter.	Length.	
		Inches.	*Inches.*	
No. 7...	*a* D1134	$1\frac{1}{2}$	42	Detonation incomplete.
No. 7...	*b* D1147	$1\frac{3}{4}$	80	Do.
No. 7...	*c* D1148	$1\frac{1}{2}$	42	Do.

a No detonators distributed in the charge.
b Three No. 7 detonators placed every one-half meter in the charge.
c Three No. 7 detonators placed every one-fourth meter in the charge.

Further tests were made with an insensitive gelatin dynamite by placing one No. 7 detonator every one-eighth meter in the charge, with results as follows:

Results of rate-of-detonation tests in which No. 7 detonators were distributed in the charge.

Grade of detonator.	Diameter of cartridges.	Dimensions of galvanized-iron tube used.		Length of charge that detonated.
		Diameter.	Length.	
	Inches.	*Inches.*	*Inches.*	*Inches.*
No. 7	1½	1¾	42	21½
No. 7	1½	1¾	42	16
No. 7	1½	1¾	42	20

The results of rate-of-detonation tests with the same explosive, when no extra detonators were used, were as follows:

Results of rate-of-detonation tests without extra detonators.

Grade of detonator.	Diameter of cartridge.	Dimensions of galvanized-iron tube used.		Length of charge that detonated.
		Diameter.	Length.	
	Inches.	*Inches.*	*Inches.*	*Inches.*
No. 7	1½	1¾	42	6
No. 7	1½	1¾	42	15
No. 7	1½	1¾	42	9

The results of the tests tabulated above indicate that extra detonators distributed 5 inches apart in a cartridge file of an insensitive explosive 40 inches long have a slight tendency to increase the propagation of the explosive wave, but that extra detonators placed 10 inches apart offer no advantage.

With an insensitive gelatin dynamite, such as that used in the tests, the influence of the detonator probably does not extend further than 5 inches. Furthermore, the explosion wave and the detonation of the explosive surrounding the detonator probably precede the explosion of the detonator. Assuming this to be true, the detonator is exploded in the products of combustion of the explosive, and, accordingly, offers little, if any, advantage as a means of extending the explosive wave.

Attention is called, however, to the fact that it is often advantageous to fire simultaneously two or more electric detonators placed in different parts of the charge in the same drill hole. Under these conditions, if the charge is fired simultaneously, the time of detonation would be reduced, and, accordingly, the shattering effect of the explosion would be materially increased. Also when long charges of explosives are used, it is sometimes necessary to use more than one electric detonator in the charge to insure complete detonation.

In quarry operations the large drill-hole method of blasting is being rapidly introduced. The former practice of quarrying by the bench method and drilling holes of small diameter, which in many cases requires the chambering of the bottom of the drill holes before loading the main charge, is more expensive.

In some quarries 6-inch holes are drilled 100 feet in depth, and several thousand pounds of explosive is used in a blast. The charges usually extend to a distance of 30 feet up from the bottom of the holes. It has been found that when one electric detonator is placed in the top of the charge it will not always produce a complete detonation of the entire charge in the drill hole; therefore two or more electric detonators distributed throughout the charge are generally used. When the most violent effect is desired in blasting, the best method of placing electric detonators in a charge 30 feet in length, irrespective of whether they are connected in series or parallel, is to place one electric detonator 5 feet from the bottom of the charge, one 5 feet below the top of the charge, and one in the center of the charge. Assuming that the entire charge detonates at a uniform rate, if the three electric detonators are fired simultaneously it can readily be seen that the duration of the explosive reaction will be one-sixth of the time that would be required if one electric detonator were used in the top of the charge.

Tests were made at the bureau's Pittsburgh testing station to determine whether simultaneous explosion would occur when four of the P. T. S. S. No. 6 electric detonators were connected in series and fired with different sources of electric current. In the tests in which a 4-hole firing machine of the dynamo-electric type was used, the time interval that elapsed between the firing of the first and the last electric detonator of each series varied from 0.0004 to 0.0050 second. As it requires only 0.0020 second for 30 feet of 40 per cent "straight" nitroglycerin dynamite to detonate, it is obvious that in many cases the only advantage in using more than one electric detonator in the same charge, when fired with a 4-hole firing machine, would be to insure complete detonation of the charge. It is to be noted that the time interval between the firing of the first and the last electric detonator is in some cases greater than the time required for 30 feet of 40 per cent dynamite to detonate. When a 4-hole firing machine is used to fire four electric detonators connected in series there is not sufficient current generated to fuse the platinum wires in the electric detonators. The wires are brought to different temperatures, depending on their cross-sectional area and, accordingly, the ignition of the priming charge in the electric detonators is not simultaneous, nor is its burning or detonation uniform. These causes are assumed to be responsible for the delay that occurs in the explosion of the electric detonators.

Further tests were made by using a 10-hole firing machine, all other conditions being the same as in the previous tests. The machine furnished ample current to fuse the platinum wires in the electric detonators and they were therefore fired practically simultaneously. The time interval was only 0.0001 second. In order to obtain the benefit of simultaneous blasting when two or more electric detonators are to be fired, the source of electric current should be such as to insure the instantaneous fusing of all the bridges in the electric detonators. This can be best accomplished in practical operations by wiring all electric detonators in parallel and using a light or power circuit for firing. If a high-pressure alternating current is the only source of electricity available, it may be necessary to install a transformer in order to obtain the proper pressure without injury to the leading wires. A lamp bank or a short length of fuse wire is sometimes placed in the electric circuit and answers the same purpose, irrespective of the kind or the pressure of the current supplied. However, if a lamp bank or a fuse wire is used, it should have a greater current-carrying capacity than is necessary to fire all of the electric detonators.

USE OF TWO KINDS OF EXPLOSIVES IN THE SAME DRILL HOLE.

In certain quarry operations in the Middle West, owing to variations in the hardness and structure of the different strata, it is necessary to use more than one kind of explosive in the same drill hole. The part of the drill hole that penetrates the hardest stratum is usually loaded with an explosive having a high rate of detonation. The remainder of the charge may be an explosive having an intermediate rate. In some cases black blasting powder is used, provided there are no pronounced clay seams or other irregularities that would allow the gases evolved on the explosion of the black blasting powder to escape before the main charge detonated. In work of this kind, the holes are drilled vertically 15 to 20 feet deep, and there is always sufficient stemming used to insure the maximum effect of the blast, even when the explosives used in the same drill hole detonate at different rates.

The practice of using combination charges of explosives has been recently adopted in some coal mines. The drill holes are usually shallow and, accordingly, do not permit the use of sufficient stemming properly to confine the gases when they are evolved at different rates. Under such conditions fires and blown-out shots are likely to result.

Several tests made at the Pittsburgh testing station to determine the energy developed by combination charges showed that there was no advantage in using them under conditions that simulated blasting in coal. In some of the tests, a No. 6 detonator (blasting cap) was inserted in the charge of dynamite, and placed in the back of the bore hole. In front of the detonator a charge of black blasting powder, containing a black powder igniter, was placed, and the free part of the drill hole was then well tamped with clay.

The results of the tests made in the ballistic pendulum, using combination charges of 40 per cent "straight" nitroglycerin dynamite and FFF black blasting powder, with and without a No. 6 detonator embedded in the explosive, were as follows:

The swings of the ballistic pendulum[a] in those tests in which the detonator was used were 3.42, 3.41, 3.40, 3.41, 3.26, 3.32, 3.01, 3.34, and 3.28 inches; average, 3.32 inches. In those tests in which no detonator was used the swings were 3.58, 3.30, 3.32, 3.38, 3.24, 3.31, 3.22, 3.36, and 3.31 inches; average, 3.34 inches.

The tests indicated that there is no advantage in using an extra detonator in the dynamite, as the explosion of the black blasting powder is sufficient to cause complete detonation. Many accidents have occurred in coal mines where combination charges containing detonators were used. When squibs are used for firing, it is necessary to insert a needle into the charge of black blasting powder, and there is always then a possibility of the needle coming in contact with the detonator.

The practice of using combination charges in coal mines offers no advantage, and, as there are many dangers attendant upon their use, the practice should be discouraged.

PUBLICATIONS ON MINE ACCIDENTS AND TESTS OF EXPLOSIVES.

The following Bureau of Mines publications may be obtained free by applying to the Director Bureau of Mines, Washington, D. C.:

BULLETIN 10. The Use of Permissible Explosives, by J. J. Rutledge and Clarence Hall. 1912. 34 pp., 5 pls.

BULLETIN 15. Investigations of Explosives Used in Coal Mines, by Clarence Hall, W. O. Snelling, and S. P. Howell, with a chapter on the natural gas used at Pittsburgh, by G. A. Burrell, and an introduction by C. E. Munroe. 1911. 197 pp., 7 pls.

BULLETIN 17. A Primer on Explosives for Coal Miners, by C. E. Munroe and Clarence Hall. 61 pp., 10 pls. Reprint of United States Geological Survey Bulletin 423.

BULLETIN 20. The Explosibility of Coal Dust, by G. S. Rice, with chapters by J. C. W. Frazer, Axel Larsen, Frank Haas, and Carl Scholz. 204 pp., 14 pls. Reprint of United States Geological Survey Bulletin 425.

BULLETIN 44. First National Mine-Safety Demonstration, Pittsburgh, Pa., October 30 and 31, 1911, by H. M. Wilson and A. H. Fay, with a chapter on the explosion at the experimental mine, by G. S. Rice. 1912. 75 pp., 7 pls.

BULLETIN 46. An Investigation of Explosion-Proof Mine Motors, by H. H. Clark. 1912. 44 pp., 6 pls.

BULLETIN 48. The Selection of Explosives Used in Engineering and Mining Operations, by Clarence Hall and S. P. Howell. 1913. 50 pp., 3 pls.

---

[a] The ballistic pendulum used by the Bureau of Mines is a large mortar swung from a pivoted support. The explosive to be tested is fired from a small cannon into the mouth of the mortar, and the swing of the mortar is taken as a measure of the strength of the explosive.

BULLETIN 52. Ignition of Mine Gases by the Filaments of Incandescent Lamps, by H. H. Clark and L. C. Ilsley. 1913. 31 pp. 6 pls.

TECHNICAL PAPER 4. The Electrical Section of the Bureau of Mines, Its Purpose and Equipment, by H. H. Clark. 1911. 12 pp.

TECHNICAL PAPER 6. .The Rate of Burning of Fuse as Influenced by Temperature and Pressure, by W. O. Snelling and W. C. Cope. 1912. 28 pp.

TECHNICAL PAPER 7. Investigations of Fuse and Miners' Squibs, by Clarence Hall and S. P. Howell. 1912. 19 pp.

TECHNICAL PAPER 11. The Use of Mice and Birds for Detecting Carbon Monoxide After Mine Fires and Explosions, by G. A. Burrell. 1912. 15 pp.

TECHNICAL PAPER 12. The Behavior of Nitroglycerin When Heated, by W. O. Snelling and C. G. Storm. 1912. 14 pp., 1 pl.

TECHNICAL PAPER 13. Gas Analysis as an Aid in Fighting Mine Fires, by G. A. Burrell and F. M. Seibert. 1912. 16 pp.

TECHNICAL PAPER 14. Apparatus for Gas-Analysis Laboratories at Coal Mines, by G. A. Burrell. 1913. 24 pp., 7 figs.

TECHNICAL PAPER 17. The Effect of Stemming on the Efficiency of Explosives, by W. O. Snelling and Clarence Hall. 1912. 20 pp.

TECHNICAL PAPER 18. Magazines and Thaw Houses for Explosives, by Clarence Hall and S. P. Howell. 1912. 34 pp., 1 pl.

TECHNICAL PAPER 19. The Factor of Safety in Mine Electrical Installations, by H. H. Clark. 1912. 14 pp.

TECHNICAL PAPER 21. The Prevention of Mine Explosions; Report and Recommendations, by Victor Watteyne, Carl Meissner, and Arthur Desborough. 12 pp. Reprint of United States Geological Survey Bulletin 369.

TECHNICAL PAPER 22. Electrical Symbols for Mine Maps, by H. H. Clark. 1912. 11 pp., 8 figs.

TECHNICAL PAPER 23. Ignition of Mine Gas by Miniature Electric Lamps, by H. H. Clark. 1912. 5 pp.

TECHNICAL PAPER 24. Mine Fires, a Preliminary Study, by G. S. Rice. 1912. 51 pp.

TECHNICAL PAPER 28. Ignition of Mine Gas by Standard Incandescent Lamps, by H. H. Clark. 1912. 6 pp.

TECHNICAL PAPER 40. Metal-Mine Accidents in the United States during the Calendar Year 1911, by A. H. Fay. 1913. 54 pp.

TECHNICAL PAPER 46. Quarry Accidents in the United States during the Calendar Year 1911, compiled by A. H. Fay. 1913. 32 pp.

TECHNICAL PAPER 48. Coal-Mine Accidents in the United States, 1896–1912, with Monthly Statistics for 1912, by F. W. Horton. 1913. 72 pp.

TECHNICAL PAPER 53. Proposed Regulations for the Drilling of Gas and Oil Wells, with Comment thereon, by O. P. Hood and A. G. Haggem. 1913. 28 pp., 2 figs.

MINERS' CIRCULAR 3. Coal-Dust Explosions, by G. S. Rice. 1911. 22 pp.

MINERS' CIRCULAR 4. The Use and Care of Mine-Rescue Breathing Apparatus, by J. W. Paul. 1911. 24 pp.

MINERS' CIRCULAR 5. Electrical Accidents in Mines; Their Causes and Prevention, by H. H. Clark, W. D. Roberts, L. C. Ilsley, and H. F. Randolph. 1911. 10 pp., 3 pls.

MINERS' CIRCULAR 6. Permissible Explosives Tested Prior to January 1, 1912, and Precautions to be Taken in Their Use, by Clarence Hall. 1912. 20 pp.

MINERS' CIRCULAR 9. Accidents from Falls of Roof and Coal, by G. S. Rice. 1912. 16 pp.

MINERS' CIRCULAR 10. Mine Fires and How to Fight Them, by J. W. Paul. 1912. 14 pp.

MINERS' CIRCULAR 11. Accidents from Mine Cars and Locomotives, by L. M. Jones. 1912. 16 pp.

○

Technical Paper 162

DEPARTMENT OF THE INTERIOR
FRANKLIN K. LANE, Secretary
BUREAU OF MINES
VAN. H. MANNING, Director

INITIAL PRIMING SUBSTANCES FOR HIGH EXPLOSIVES

BY

GUY B. TAYLOR

AND

W. C. COPE

WASHINGTON
GOVERNMENT PRINTING OFFICE
1917

The Bureau of Mines, in carrying out one of the provisions of its organic act—to disseminate information concerning investigations made—prints a limited free edition of each of its publications.

When this edition is exhausted, copies may be obtained at cost price only through the Superintendent of Documents, Government Printing Office, Washington, D. C.

The Superintendent of Documents *is not an official of the Bureau of Mines.* His is an entirely separate office and he should be addressed:

SUPERINTENDENT OF DOCUMENTS,
Government Printing Office,
Washington, D. C.

The general law under which publications are distributed prohibits the giving of more than one copy of a publication to one person. The price of this publication is 5 cents.

First edition. November, 1917.

CONTENTS.

TABLES.

INITIAL PRIMING SUBSTANCES FOR HIGH EXPLOSIVES.

By GUY B. TAYLOR and W. C. COPE.

INTRODUCTION.

In order to insure effective action by high explosives, or explosives of the second order,[a] it is necessary to use a detonator. As most of the technically useful blasting explosives and all the permissible explosives used in coal mines are explosives of the second order, an exact knowledge of the theory and practice of initiating their explosion is highly desirable. Misfires frequently cause accidents, and many misfires and incomplete detonations of explosives result from the failure or the ineffectiveness of the detonator. The Bureau of Mines is frequently asked to investigate such difficulties and is constantly testing and analyzing detonators. Besides making these practical tests, the bureau is to investigate the properties of initiating substances.

The course that an explosive reaction may take is influenced by the character of the initial impulse imparted to the explosive and also by the nature of the explosive itself. In other words, the effect produced by an explosive in practice depends not only on the explosive, but also on the primer [a] used to set it off.

Practically all detonator compositions used in this country have as their essential ingredient fulminate of mercury. Heretofore, the most common method of varying the character of the primer has been that of varying the weight of the fulminate mixture in the copper shell, that is, by using stronger or weaker detonators. Recent investigations have shown that there are primers more efficient than mercury fulminate mixtures.

---

[a] In this report the author considers explosives as of two orders. First order explosives are of the slow-burning type, such as black powder. Second order explosives are of the detonating type, such as dynamite.

[a] The word " primer " in this report signifies any substance which by burning or explosion in contact with the principal explosive charge causes the latter to explode. The primer used for blasting explosives used in mines is called a detonator or blasting cap.

DEVELOPMENT OF PRIMERS FOR HIGH EXPLOSIVES.

" Initial ignition," says Brunswig (12),[a] "or, more generally, initial impulse, is that impulse which is necessary in order that a sensitive system, capable of exothermal transformation, is caused to explode. Since the resistance which, especially in the case of heterogeneous systems, prevents spontaneous explosive decomposition must first be overcome, the initial impulse performs a very essential though only preparatory work."

For explosives of the first order, such as black powder, the initial impulse is supplied by the momentary application at a single point of a source of high temperature, usually the flame from a fuse, a match, or a percussion cap. Explosives of the second order require stronger means of ignition, and the initial impulse is usually provided by exploding a first-order, highly brisant explosive in contact with them. Initial priming explosives may be defined as explosives of the first order that when exploded in small quantities will transmit detonation to high explosives. This definition excludes intermediate priming explosives, the so-called " boosters," as a booster fired as a primary charge, at least in small quantities, has no power of initiating detonation.[a]

Nitroglycerin, discovered by Sobrero in 1846, was not industrially applied until the early sixties. Nobel, to whose indomitable perseverance the world owes the development of this agent now so important in engineering undertakings, discovered the means of detonating it surely and effectively by mercury fulminate in 1864. Previous to that time Nobel (1) had experimented with mixtures of black powder and nitroglycerin inclosed in wooden or metallic cartridges. These relatively large charges fired under the liquid nitroglycerin by means of a fuse were not wholly satisfactory. The modern use of high explosives begins with Nobel's introduction of dynamite—nitroglycerin absorbed by a porous material—in 1867 and the small drawn copper or zinc shell filled with mercury fulminate (5).

At first the fulminate was mixed with black powder, then with potassium nitrate, and later with potassium chlorate. The chlorate mixture is to-day practically the only priming composition of commercial importance.

In 1888 a patent (101) granted to Nobel proposed to substitute a mixture of potassium chlorate and lead picrate, but even this

---

[a] The numbers refer to those printed in the bibliography at the end of this paper.

[a] Dry guncotton, long used as an intermediate primer for wet guncotton, must be fired by an initial primer, such as fulminate. Guncotton is both a primary and a secondary explosive, according to the means of ignition, but has no priming action if set off by means of a fuse. Loose-powdered nitro compounds are used as intermediate primers for compressed or cast nitro compounds.

mixture required a superimposed primer of mercury fulminate in order to assure the maximum effect. The next advance was the proposal of Wöhler (114), in 1900, to apply the principle of the booster charge by substituting nitro compounds for part of the fulminate in blasting caps.

Following the work of Curtius (42) on hydronitric acid Will and Lenze (30) experimented with hydronitrides (azides) at the military testing station at Spandau. A fatal accident terminated these experiments and they were kept secret by the German war office. Wöhler (33) in seeking for primers to replace fulminate, and in ignorance of the earlier work of Will and Lenze, discovered the great effectiveness of hydronitrides and published his work in 1907. Since that time lead hydronitride (or azide, PbN_6) has been put on the market by a German concern, but has not been manufactured extensively in this country, whether for technical or commercial reasons is not evident. Recently patents (91, 94) have been issued for primers consisting of pure organic compounds without any metal. It is claimed that bis-diazonitrobenzene perchlorate is superior to fulminate and even the azides as an initiating explosive (23).

THEORIES OF ACTION OF INITIAL PRIMERS.

The wave theory of detonation in high explosives as developed by Berthelot (10) finds general acceptance to-day. There have been two main theories of the mechanism of the transmission of the explosion wave from the initiating explosive to the high explosive. Abel (6) advanced the hypothesis of wave synchronism. His experiments showed that the highly brisant explosives, nitrogen chloride and nitrogen iodide, have no priming effect on nitroglycerin or guncotton. Only once did 3.5 grams of nitrogen chloride detonate guncotton, whereas one-tenth as much mercury fulminate invariably did so. Abel ascribed the peculiar action of the fulminate to a synchronism between the explosion wave of the fulminate and that of the high explosive. Threlfall (29) in criticizing the Abel hypothesis suggested a theory of vortex rings. Meyer (20) holds that, according to Abel's hypothesis, every explosive ought to be its own best detonating primer.

The second theory strives to account for the action of initial primers as a mechanical and heat effect. This view seems to have been held by Nobel as well as by Berthelot. Its latest champion is Wöhler (31). His idea, in brief, is that the effect of initial priming results from enormous momentary pressure. In order to obtain this pressure, a rapid rate of decomposition accompanied by high temperature is prerequisite. Any mixture of substances, or any combination of substances that, when placed in the detonator shell and ignited by flame or spark, detonates with such velocity that the walls of the shell do not yield until the entire gaseous product has so accumulated that it can escape all at once, and deliver by its kinetic energy a colossal blow, will be an efficient primer. Storm and Cope (27) consider the initial rate of detonation as the most important property of primers, and state that this initial rate is far higher than is indicated by the methods used for measuring rate of detonation of blasting explosives. Their views are not at variance with those of Wöhler. Stettbacher (23) is not so ready to abandon entirely the wave-synchronism theory and considers that Wöhler leaves much unexplained.

WORK OF OTHER INVESTIGATORS.

Wöhler and Matter (321) investigated the explosive properties and the priming effects of seven explosives besides mercury fulminate. Tests were made to ascertain their expansion in lead blocks, their penetrating effect on lead plates, and the temperature at which they ignited. The results are given in Table 1.

TABLE 1.—*Results of tests by Wöhler and Matter to ascertain explosive effects of initial primers.*

Primer.	Lead-block test.		Lead-plate test.		Ignition temperature.
	Loading density (maximum).	Chamber made by 2-gram charge.	Weight required for complete perforation.	Relative order of general effect.	
	Grams				
	c. c.	*C. c.*	*Grams.*		° *C.*
Silver hydronitride a	3.382	22.6	0.10	1	290
Mercury fulminate	2.112	25.6	.20	2	190
Trimercury aldehyde perchlorate		18.3		3	
Trimercury aldehyde chlorate	2.995	15.3	.40	4	130
Diazobenzene-nitrate	1.459	43.1	1.5 to 2.0	5	90
Nitrogen sulphide	2.112	39.2	1.5 to 2.0	6	190
Sodium fulminate	1.651	14.9	No perforation.	7	150
Basic mercury nitromethane	2.492	7.5	No perforation.	8	160

a The salts of the acid (HN_3) are variously known as hydronitrides, trinitrides, nitrides, and azides.

To determine the actual priming effect of these same substances, Wöhler weighed 1 gram of a nitro compound into a No. 10 copper detonator shell, and its surface was smoothed by slight pressure, and the required quantity of the priming being tested was added on top. A small perforated brass capsule was placed on top of the priming and a pressure of 2,000 kilograms per square centimeter was applied. For the mercury aldehydes a pressure of only 400 kilograms per square centimeter was used. By firing these loaded detonator shells on lead plates, the minimum quantity of priming required for the complete detonation of the nitro compound was determined. The results are given in Table 2 following:

TABLE 2.—*Results of tests conducted by Wöhler to determine minimum weight of primer required to detonate nitro compounds.*

[Results expressed in grams.]

Primer.	Gun-cotton.	Picric acid.	Tri-nitro re-sorcin.	Tri-nitro cresol.	Tri-nitro benzoic acid.	Tri-nitro ben-zene.	Tri-nitro xylene.	Tri-nitro to-luene.
Silver hydronitride	0.05	0.027	0.08	0.05	0.20	0.05	0.25	0.05
Mercury fulminate	.20	.25	.20	.30	.25	.30	.30	.30
Trimercury aldehyde perchlorate		.15	(a)	.40		.40		
Trimercury aldehyde chlorate	.30	.50			(b)			
Diazobenzene nitrate		(c)						
Nitrogen sulphide		(c)						
Sodium fulminate		(c)						
Mercury nitro methane		(c)						

a 0.40-gram charge failed.   b 0.50-gram charge caused partial detonation.   c 0.50-gram charge failed.

In a similar manner Martin (19) determined the minimum charge for some fulminates and azides. One-half gram of the nitrocompound was used, the loading pressure being 1,100 kilograms per square centimeter. Martin introduced the conception of work density (A_d) derived from the formula $A_d = \dfrac{1.033\ V_o\ T}{273 \times 100}$ (kg. m. per c. c.), in which V_o is the volume of gas at 0° C. and 1 atmosphere evolved from 1 c. c. of the explosive, and T is the absolute explosion temperature calculated from the heat of explosion. The results of tests recorded in Table 3 following indicate that there is a lack of parallelism between "work density" and priming efficiency.

TABLE 3.—*Results of tests by Martin to determine minimum weight of primer required to detonate nitrocompounds.*

[Results in grams.]

Primer.	Tetra-nitro-methyl-anilin.	Picric acid.	Trinitro-toluene.	Trinitro-anisol.	Work density. (A_d).	Loading density. (Grams c. c.)
Mercury fulminate	0.29	0.30	0.36	0.37	152.15	3.298
Silver fulminate	.02	.05	.095	.23	101.0	3.2
Cadmium fulminate	.008	.05	.11	.26	152.0	3.002
Mercurous azide	.045	.075	.145	.55	110.4	3.78
Silver azide	.02	.035	.07	.26	95.95	2.98
Lead azide	.025	.025	.09	.28	98.96	3.01
Cadmium azide	.01	.02	.04	.10	119.3	2.20

In a report issued by the Rheinisch-westfälische Sprengstoffe Aktien-Gesellschaft (28) comparisons were made of the effects of "straight" mercury fulminate, of its mixture with 20 per cent of potassium chlorate, and of lead azide (PbN_6). All three gave about the same effect in perforating lead plates. The chlorate mixture hollowed out a cavity of 33.5 c. c. in a lead block; the straight fulminate produced a cavity of 25.3 c. c. and the lead azide a cavity of 21.5 c. c. The detonating efficiencies on 1 gram of nitrocompound tested in a copper shell 45 by 6.85 mm. with the inner perforated capsule (so-called reinforced detonator) were as follows:

Results of tests to determine detonating efficiencies of three primers.

Primer.	Quantity required to detonate—	
	Trinitro-toluene.	Tetranitro-methyl-anilin.
	Grams.	*Grams.*
Pure fulminate.........................	0.35	0.40
Mixture of 80 per cent mercury fulminate and 20 per cent potassium chlorate.............................	.30	.20
Lead azide.............................	.08	.01

Storm and Cope (27) have used the "sand test" to determine the efficiencies of initial priming compositions, particularly mixtures of mercury fulminate and potassium chlorate.. For the details of this elaborate series of tests the original paper should be consulted. The results showed that as to ability both to crush quartz sand and to detonate nitrosubstitution compounds the compositions stood in the following order: Eighty per cent fulminate, 20 per cent chlorate; 90 per cent fulminate, 10 per cent chlorate; straight fulminate.

RESULTS OF EXPERIMENTS.

The "sand test" as developed by Storm and Cope (27) offers an unusually good device for the laboratory study of initial priming explosives. In a previous report[a] two explosives of the second order, trinitrotoluene (TNT) and tetranitromethylanilin (tetryl) were shown to be suitable substances for testing the efficiency of initial priming explosives, and the method of using them was indicated. This report gives the results of tests with several substances that have been proposed as initial primers as well as with new mixtures.

MERCURY FULMINATE WITH AN OXYGEN CARRIER.

For various reasons mercury fulminate seems destined for some time to come to hold its place as the essential ingredient in high explosive primers. At the present time (1915) straight mercury fulminate is used in France (72), and mixtures of fulminate with potassium chlorate in all other countries, for charging detonators. The admixture of potassium chlorate not only reduces the quantity of

---

[a] Taylor, G. B., and Cope, W. C., Sensitiveness to denonation of trinitrotoluene and tetranitromethylanilin. Tech. Paper 145, Bureau of Mines, 1916, 13 pp.

the more expensive mercury fulminate but makes a composition running much more easily in the loading machines. According to the tests of Storm and Cope mixtures containing up to 20 per cent potassium chlorate are more efficient than the fulminate alone. The explosive reaction by which mercury fulminate decomposes may be written as follows:

$$HgC_2N_2O_2 = Hg + 2CO + N_2$$

Addition of an oxygen carrier, such as chlorate, oxidizes the CO to CO_2, thus increasing the total energy of the explosion. Potassium chlorate itself decomposes exothermically and adds additional heat to the reaction. The endothermic character of this compound is often advanced as a reason for its use, but, as shown subsequently, many compounds having a decidedly positive heat of formation are equal to or better than potassium chlorate as a component of detonator compositions containing mercury fulminate.

METHOD OF WORK.

Mixtures of mercury fulminate with 10 and 20 per cent of the oxygen carrier were tested by determining the minimum weights of them necessary to cause certain detonation of TNT and tetryl. The mercury fulminate used was all taken from the same lot of No. 7 detonators (M–54), the $KClO_3$ being washed out thoroughly with water and the fulminate being dried over night in a Freas oven at 50° C. The sample was very fine grained. The oxygen carriers (salt, oxide, etc.) were chemically pure preparations ground in an agate mortar and were always used in a thoroughly dry condition. The mixtures were prepared in 3-gram to 6-gram quantities by mixing intimately the accurately weighed components with a camel's-hair brush on a sheet of glazed paper. The $KClO_3$ mixtures were taken from commercial detonators that were shown by analysis to be of satisfactory purity and composition.

The nitrocompound (0.40 gram) was placed in a drawn-copper detonator shell 5.5 mm. ($\frac{7}{32}$ inch) in inside diameter and 38 mm. (1½ inches) long, and pressed down lightly with a glass rod. Varying weights of the priming composition were next placed on top of the nitrocompound and a closely fitting perforated inner copper capsule (reinforcing cap) 9 mm. long with 2.3 mm. perforation was pressed down over the charge under a loading pressure of 100 pounds (200 atmospheres) per square inch.[a] A short piece of fuse was crimped into the shell and the detonator was fired in the sand bomb. The sand test showed conclusively by the weight of sand crushed the degree of detonation of the nitrocompound. Thus the mini-

---

[a] Loading machine for pressing caps is described by Storm, C. G., and Cope, W. C., The sand test for determining the strength of detonators: Tech. Paper 125, Bureau of Mines, 1916, p. 48.

mum weight of the primer necessary to insure complete detonation was determined and the efficiencies of the various priming mixtures were compared.

Three to five (generally five) complete detonations of the nitrocompound and no failures were obtained for each composition at the minimum charge. In all about a thousand " sand tests" were made in the course of the investigation. It will not be necessary to give the complete details. The dividing line between " complete " and " partial " detonation was as a rule sharp and it was possible to determine within 0.01 gram the weight of primer necessary to detonate completely the TNT or tetryl. Strong detonations often occurred with charges several hundredths of a gram less than the established minimum charge, but the criterion of minimum charge was that of certainty of detonation.

TESTS WITH OXYHALOGEN SALTS.

Ammonium perchlorate was suggested as a substitute for potassium chlorate by Alvisi (76), and potassium bromate has also been patented (100) for use with mercury fulminate. Neither seems to have had any extended commercial application, though both make efficient mixtures. The efficiencies for detonating TNT (M–1789) and tetryl (M–1849) are given in Table 4. Under the same conditions of loading the minimum charge of straight mercury fulminate is 0.26 gram for TNT and 0.24 gram for tetryl. It will be noted that all the oxyhalogen salt mixtures, with the possible exception of potassium iodate, were as effective as the straight fulminate or more so. The heats of formation are given from the corresponding halide and oxygen. The two exothermic salts were slightly less effective than the endothermic salts.

TABLE 4.—*Results of tests to determine minimum weights of mixtures of mercury fulminate and oxyhalogen salts required to detonate TNT and tetryl reinforced detonators.*

[Reinforcing cap pressed over each charge at a pressure of 200 atmospheres per square inch.]

Oyhalogen salt mixed with mercury fulminate.	Quantity required to detonate TNT having a proportion of oxygen carrier of—		Quantity required to detonate tetryl having a proportion of oxygen carrier of—		Heat of formation.
	10 per cent.	20 per cent.	10 per cent.	20 per cent.	
	Grams.	*Grams.*	*Grams.*	*Grams.*	*Kilogram calories.*
KClO₃	0.25	0.24	0.19	0.17	−11.9
KClO₄	.24	.26	.20	.20	7.9
KBrO₃	.21	.21	.16	.16	−11.3
KIO³	.26	.28	.24	.22	45.9
KIO₄	.21	.22	.18	.19	....
Ba(ClO₃)₂.H₂O	.21	.22	.17	.18	*a* −10.9
NaClO₃	.22	.22	.17	.17	−13.1
NH₄ClO₄	.23	.25	.18	.20	....

a Calculated as being formed from oxygen, steam, and BaCl₂.

TESTS WITH NITRATES AND OTHER OXIDIZERS.

Table 5 gives the minimum charges of mixtures of mercury fulminate with nitrates, lead oxides, lead chromate, potassium bichromate, and potassium permanganate. It will be noted that the silver and lead nitrates make efficient mixtures, whereas the alkali and alkaline earth nitrates do not. All of the compounds represented in this table absorbed heat in their decomposition, yet 7 of the 11 make mixtures with mercury fulminate having greater priming efficiency than the fulminate alone. The quantity of heat subtracted or added to the total heat of explosion by the decomposition of the oxygen carrier is never more than a few per cent. Besides the substances listed in the tables, the following were also tried in 10 and 20 per cent mixtures: Manganese dioxide (MnO_2), yellow mercuric oxide (HgO), the uranium oxides (UO_3 and U_3O_8), cupric oxide (CuO), vanadium oxide (V_2O_5), ferric oxide (Fe_2O_3), and finely divided platinum and platinum black. These mixtures were much less efficient than mercury fulminate alone, most of them having minimum charges of more than 0.40 gram.

TABLE 5.—*Results of tests to determine minimum weights of mixtures of mercury fulminate and various oxygen carriers required to detonate TNT and tetryl reinforced detonators.*

[Reinforcing cap pressed over each charge at a pressure of 200 atmospheres per square inch.]

Oxygen carrier mixed with mercury fulminate.	Quantity required to detonate TNT having a proportion of oxygen carrier of—		Quantity required to detonate tetryl having a proportion of oxygen carrier of—	
	10 per cent.	20 per cent.	10 per cent.	20 per cent.
	Grams.	*Grams.*	*Grams.*	*Grams.*
AgNO$_3$	0.26	0.26	0.21	0.21
Pb(NO$_3$)$_2$	.24	.22	.17	.18
KNO$_3$	.28	+.40	.22	+.40
NaNO$_3$	.28	+.40	.24	+.40
Ba(NO$_3$)$_2$	.28	.32	.24	.24
PbO	.23	.26	.20	.22
Pb$_3$O$_4$	.22	.28	.17	.24
PbO$_2$	.22	.27	.17	.20
PbCrO$_4$	.26	+.40	.22	+.36
K$_2$Cr$_2$O$_7$	.24	.30	.20	.26
KMnO$_4$	.22	.24	.16	.18

CONCLUSIONS REGARDING EFFICIENCIES OF FULMINATE MIXTURES.

The results of the experiments with priming compositions consisting of mixtures of mercury fulminate and various oxygen carriers do not disclose any striking conclusion regarding the relation between any single property of the oxidizer and the priming efficiency of the mixture. Oxidizers varying widely in specific gravity, heat of formation, percentage of available oxygen, character of the nonvolatile product of explosion, and general chemical relationship yield about equally efficient mixtures. Although it has not been found possible to trace any general relation of specific properties to effects, the following facts may be pointed out: Salts of the oxyhalogen acids make efficient mixtures; oxides and nitrates of heavy metals whose decomposition results in free metal make especially effective mixtures; heavy metal oxides that reduce to lower oxides are ineffective.

TESTS WITH LEAD HYDRONITRIDE, SILVER ACETYLIDE, AND THEIR MIXTURES WITH MERCURY FULMINATE.

Lead hydronitride or lead azide (PbN_6) has been shown by several investigators to be an effective initial primer (16, 19, 28, 32). It is said that its mixtures with mercury fulminate do not become "dead pressed" (102). Recently Stettbacher (26) has prepared and tested silver acetylides (107) precipitated from ammoniacal, neutral, and nitric acid solutions by purified acetylene gas. He found that the precipitates from the neutral and the acid solutions were probably identical and did not correspond to the formula Ag_2C_2, but their yield and explosive power indicated that they contained some oxygen. The "limit charge" determined for tetryl was 0.07 gram.

For tests represented in Table 6 following, silver acetylide was prepared according to Stettbacher's method from acid and neutral solutions, but no minimum charge could be established up to 0.40 gram. Admixed with mercury fulminate, however, it added considerably to the efficiency of the fulminate. The composition of silver acetylide is probably not constant and the product tried may have been different from Stettbacher's preparation. The precipitates from the acid and the neutral solutions had about the same properties.

The lead hydronitride used in the tests was prepared by precipitation from a lead acetate solution with a 4 per cent solution of sodium hydronitride made slightly acid with acetic acid.

TABLE 6.—*Minimum weight, in grams, of murcury fulminate, lead hydronitride, and silver acetylide in reinforced detonator shells pressed to 200 and 400 atmospheres per square inch required to detonate TNT and tetryl.*

Primer.	Loading pressure of 200 atmospheres.		Loading pressure of 400 atmospheres.	
	TNT.	Tetryl.	TNT.	Tetryl.
Mercury fulminate	0.26	0.24	0.30	0.25
Lead hydronitride	.23	.01	.14	.04
Silver acetylide	.40+	.40+	.40+	.40+
Mixture of 80 per cent of mercury fulminate and 20 per cent of lead hydronitride	.18	.06	.14	.06
Mixture of 80 per cent of mercury fulminate and 20 per cent of silver acetylide	.16	.10	.18	.10

The results show that when mercury fulminate[a] is used as a priming, especially for TNT, the minimum weight required for certain detonation is greater when the charge is pressed under a pressure of 400 atmospheres than when pressed under a pressure of 200 atmospheres per square inch. Exactly the opposite is true of lead hydronitride and its mixtures with mercury fulminate.[b] In seeking an explanation, certain experiments were conducted. The results are shown in Tables 7 and 8. The TNT was first loaded into the detonator shell and pressed to the desired pressure. Careful determinations of the density of loading at each pressure were made by cutting the shell apart and waterproofing the exposed surface with a thin collodion film and weighing in a pycnometer. From the known weights of TNT, copper, and water displaced, the density was calculated. The densities ascertained as compared with water at 27° C., were 1.305 when the TNT was pressed under a pressure of 200 atmospheres, and 1.387 when it was pressed under a pressure of 400 atmospheres per square inch.

The reinforcing capsule ordinarily used in this work was long enough to inclose all of the priming and a little of the TNT. In order to avoid disturbance of the compressed TNT, a shortened capsule was used in the experiments representd in Table 7. It will be noted that this change in the construction of the detonator considerably affected the size of the minimum charge.

The results presented in Tables 6, 7, and 8 indicate that uncompressed mercury fulminate is more efficient than compressed fulminate, and that the efficiency probably decreases with increase in pressure. It seems also that the sensitiveness of TNT to detonation probably decreases as the degree of compression increases. The efficiency of primers containing lead hydronitride probably increases with the pressure of loading, and this effect is probably strong enough to overcome the slight decrease in sensitiveness of the TNT.

---

[a] It has been previously shown also for chlorate-fulminate mixtures that increasing the pressure of loading the detonator decreases the probability of detonation of TNT. See Storm, C. G., and Cope, W. C., The sand test for determining the strength of detonators: Tech. Paper 125, Bureau of Mines, 1916, pp. 49–53.

[b] No particular significance should be attached to the apparent increase in minimum charge of lead hydronitride for tetryl that has been loaded at a pressure of 400 atmospheres over that loaded at a pressure of 200 atmospheres. With the type of reinforcing capsule used all of the primer could not be kept properly under the cap during the pressing at this pressure. In general, the results were always more erratic at the higher pressure for all primers tested.

TABLE 7.—*Results of sand tests of detonators loaded with 0.40 gram of TNT at different pressures, each primer at a constant pressure of 200 atmospheres per square inch; shortened reinforcing capsules.*

Loading pressure of TNT, atmospheres per square inch.	Weight of sand pulverized finer than 30-mesh with priming charge of—												
	Straight mercury fulminate.									Mixture of 80 per cent mercury fulminate and 20 per cent lead hydronitride.			
	0.46 gram.	0.44 gram.	0.42 gram.	0.40 gram.	0.38 gram.	0.36 gram.	0.34 gram.	0.32 gram.	0.30 gram.	0.20 gram.	0.18 gram.	0.16 gram.	0.14 gram.
	Grams.	*Grams.*	*Grams.*	*Grams.*	*Grams.*	*Grams.*	*Grams.*	*Grams.*	*Grams.*	*Grams.*	*Grams.*	*Grams.*	*Grams.*
200.....	......	......	38.0	35.8	37.1	35.8	31.8	a17.6	a7.3	33.0	29.8	30.5	a12.0
	......	......	37.1	37.7	a13.3	a23.0	a10.3	......	......	......	29.2	a25.2	......
	......	......	36.7	a14.1	......	......	......	......	......	......	29.4	......	......
400.....	38.0	36.3	36.5	a14.5	......	......	a11.8	......	a9.2	......	29.5	30.6	a14.4
	40.5	a18.7	a21.5	a16.0	......	......	......	......	......	......	28.5	a25.5	......
	38.1	......	......	......	......	......	......	......	......	......	29.0	......	......

a Incomplete detonation.

TABLE 8.—*Results of sand tests of detonators loaded with 0.40 gram of TNT at different pressures and primed with compressed and uncompressed mercury fulminate, without reinforcing capsules.*

Loading pressure of TNT, atmospheres per square inch.	Compression of primer per square inch.	Weight of sand pulverized finer than 30-mesh by priming charge of—					
		0.425 gram.	0.400 gram.	0.375 gram.	0.350 gram.	0.325 gram.	0.300 gram.
200	Uncompressed...............	......	......	......	31.2	a10.0	a9.9
		......	......	......	34.8	......	......
		......	......	......	28.5	......	......
	200 atmospheres...............	......	37.2	34.0	a14.0	......	......
		......	34.7	a11.4	......	......	......
		......	40.5	......	......	......	......
400	Uncompressed...............	......	......	37.7	a18.1	......	......
		......	......	38.3	......	......	......
		......	......	38.2	......	......	......
	200 atmospheres...............	36.8	a19.2	......	......	......	......
		33.4	......	......	......	......	......
		32.3	......	......	......	......	......

a Incomplete detonation.

ACCELERATING ONE PRIMING BY ANOTHER.

That the action of the less efficient priming explosives, including mercury fulminate, may be considerably strengthened by superimposing a quicker acting explosive upon them has been suggested by Wöhler (31). The efficiency of lead hydronitride and of silver acetylide when used in this way is shown by the results presented in Table 9. The acetylide, although an inefficient priming for TNT or tetryl, markedly increases the effectiveness of mercury fulminate. The probable reason for this anomaly is discussed subsequently.

TABLE 9.—*Results of sand tests of detonators loaded with 0.40 gram of TNT and various primers, showing effect of 0.02 gram of accelerating priming.*

Loading pressure per square inch.	0.02 gram accelerating primer.	Main primer.	Weight of sand pulverized finer than 30 mesh with main priming charge of—												
			0.32 gram.	0.30 gram.	0.28 gram.	0.26 gram.	0.24 gram.	0.22 gram.	0.20 gram.	0.18 gram.	0.16 gram.	0.14 gram.	0.12 gram.	0.10 gram.	0.08 gram.
Atmospheres.															
200		Mercury fulminate	35.4	34.2	34.7	33.0, 32.7	a12.6								
200	Lead hydronitride	do									30.7, 29.5	27.0, 29.1, 25.8, 28.2	27.2, 26.2, 26.4, 26.3	a22.7	
200	Silver acetylide	do							30.7		26.6			24.4	a13.1
400		do		32.2, 31.9	30.7, 26.9	a8.8	a6.8								
400	Lead hydronitride	do									28.9				
200	Mercury fulminate	Silver acetylide		a7.4, a8.2					a5.6, a5.5		a6.8				
200	Lead hydronitride	do											27.0, 30.0, 29.5	28.0, 28.0, 30.0	25.5, a14.9

a Incomplete detonation.

The use of organic explosive compounds has frequently been suggested. In order to indicate the possibilities in that direction, two compounds were prepared.

Nitrodiazobenzene perchlorate (51) was prepared by dissolving 3.5 grams of meta nitraniline in 100 c. c. of water containing 4 c. c. of concentrated hydrochloric acid, cooling the solution with a freezing mixture, and diazotizing it with a concentrated solution of sodium nitrite. After the liquid had been filtered, a concentrated solution of ammonium perchlorate was added, and the precipitated nitrodiazobenzene perchlorate was collected on a filter paper and dried in a desiccator over sulphuric acid.

The compound was lightly pressed on top of 0.40 gram of both TNT and tetryl, without the reinforcing capsule. It was found that 0.10 gram was sufficient to detonate completely TNT, and 0.05 gram caused partial detonation. As little as 0.05 gram caused tetryl to detonate. As the experimenter was attempting to press some of the compound under the reinforcing capsule on top of tetryl, the detonator exploded violently, wrecking the press block. When pressed in lightly by hand, 0.5275 gram of the compound nearly filled a No. 8 5.5-mm. detonator shell, and the same quantity pulverized 46.6 grams of sand finer than 30 mesh in the sand bomb. A small quantity when struck with a hammer on an anvil or touched with a match in the open exploded with a loud report.

Hexamethylene-triperoxyd-diamine was prepared by the method of Girsewald (50, 91); 10.5 grams of citric acid was dissolved in 35 grams of 30 per cent hydrogen peroxide solution, and 7 grams of hexamethylene-tetramine was added. The hexamethylene-triperoxyd-diamine separated out as small white crystals. After the solution had stood 3 hours these crystals were filtered off and dried in a desiccator. A small portion was dried satisfactorily in an oven at 70° C.

This compound was tested for its ability to detonate TNT and tetryl. The test was similar to that described above except that the compound was separated in the detonator shell from the base charge by thin disks of tin foil; 0.15 gram caused TNT to detonate completely, and 0.10 gram caused nearly complete detonation; tetryl detonated completely when primed with 0.10 gram and partly when primed with 0.05 gram. The ability of the compound to crush sand (0.34 gram crushed 26.3 grams of sand finer than 30 mesh) is about the same as that of nitrodiazo-benzene perchlorate and much greater than that of mercury fulminate. When struck with a hammer on an anvil the compound explodes with a loud report, but when a match is applied to a small quantity in the open, it explodes with little noise.

The priming efficiency of these two organic explosives is high. Their loading density is much less than that of the priming compounds containing a metal, but weight for weight they are much more efficient. Under the same conditions of experiment at least 0.30 gram of lead hydronitride and 0.35 gram of mercury fulminate are required to detonate TNT. The presence of a heavy metal is not a requisite for initial priming explosives.

INTERPRETATION OF RESULTS OF THE SAND TEST.

Storm and Cope (27) found that mixtures of mercury fulminate with potassium chlorate crushed more sand finer than 30 mesh in the "sand bomb" than did mercury fulminate alone. In order to determine whether this ability to break up quartz sand always paralleled the priming efficiency (the measure of priming efficiency being taken as the relative minimum weights required to cause detonation), sand tests with 1 gram of different priming compositions were made in three ways, as follows: (a) The shell was loaded without pressure, being merely tapped on the table top; (b) The contents of the loaded shell were subjected to a pressure of 200 atmospheres per square inch; and (c) the detonator shell was "reinforced" and the contents pressed at 200 atmospheres per square inch. After each test the total quantity of sand passing through a 30-mesh screen was determined and to measure the degree of fineness, that passing a 100-mesh screen was also observed. The results for all mixtures tested are given in Table 10. They do not show any parellelism between quantity of sand crushed and priming efficiency. The results of tests showing that a pressure of loading up to 400 atmospheres has no material effect on the quantity of sand crushed are given in Table 11.

TABLE 10.—*Results of comparative sand tests of priming compositions.*

[1 gram of primer used.]

Primer mixed with mercury fulminate.	Weight of sand pulverized.						Minimum charge in grams for TNT 200 atmospheres reinforced.
	Shell not pressed.		Shell pressed under pressure of 200 atmospheres per square inch.		Detonator reinforced and pressed under a pressure of 200 atmospheres per square inch.		
	30-mesh.	100-mesh.	30-mesh.	100-mesh.	30-mesh.	100-mesh.	
	Grams.	*Grams.*	*Grams.*	*Grams.*	*Grams.*	*Grams.*	
Mercury fulminate	36.10	23.35	35.30	21.80	33.60	20.45	0.26
KClO₃, 10 per cent	38.75	25.15	36.95	23.60	36.70	22.70	.25
KClO₃, 20 per cent	39.90	25.50	39.25	24.80	38.95	22.85	.24
KClO₄, 10 per cent	39.30	25.70	38.40	24.50	36.90	24.30	.24
KClO₄, 20 per cent	41.00	27.00	41.20	26.10	38.05	24.90	.26
Ba(ClO₃)₂.H₂O, 10 per cent	39.10	24.10	36.80	23.50	35.20	23.40	.21
Ba(ClO₃)₂.H₂O, 20 per cent	39.35	25.05	38.70	24.80	36.60	24.00	.22
Pb(NO₃)₂, 10 per cent	36.50	23.35	36.35	23.50	34.30	22.55	.24
Pb(NO₃)₂, 20 per cent	37.40	23.70	35.60	22.30	34.15	22.20	.22
NaNO₃, 10 per cent	38.05	23.70	38.80	23.75	35.85	22.85	.28
NaNO₃, 20 per cent	38.90	23.35	37.55	21.15	35.75	22.60	>.40
PbO₂, 10 per cent	34.40	22.40	33.45	21.50	32.70	20.35	.22
PbO₂, 20 per cent	32.80	20.90	32.20	20.00	31.30	18.95	.27
KMnO₄, 10 per cent	34.90	22.00	34.60	21.50	32.70	19.20	.22
KMnO₄, 20 per cent	34.40	21.30	34.00	21.30	31.90	20.25	.24
MnO₂, 10 per cent	32.40	18.60	11.10	2.5	31.30	18.20	>.40
PbN₆, 20 per cent	37.20	22.60	36.7	22.3			.18
Silver acetylide, 20 per cent	34.65	20.2					.16
KClO₃, 10 per cent *a*	41.35	25.25	38.90	23.45	38.15	23.50	.12
KClO₃, 10 per cent *b*	39.45	24.10	39.45	22.95	36.10	23.45	.25

a 0.02 gram of PbN₆ superimposed.
b 1.02 grams of primer for comparison with results of tests with PbN₆ superimposed.

TABLE 11.—*Results of sand tests of 1-gram charges of priming explosives at different pressures of loading.*

Primer.	Weight of sand pulverized.					
	Shell uncompressed.		Loading pressure of 200 atmospheres per square inch.		Loading pressure of 400 atmospheres per square inch.	
	30-mesh.	100-mesh.	30-mesh.	100-mesh.	30-mesh.	100-mesh.
	Grams.	*Grams.*	*Grams.*	*Grams.*	*Grams.*	*Grams.*
Mercury fulminate	36.10	23.35	35.30	21.80	36.90	21.45
Lead hydronitride	35.20	21.30	34.45	20.60	34.75	21.00
Mixture of 80 per cent mercury fulminate and 20 per cent lead hydronitride	37.20	22.60	36.70	22.30	35.85	21.15
Silver acetylide	23.00	12.30	21.30	10.45	22.00	11.70
Mixture of 80 per cent mercury fulminate and 20 per cent silver acetylide	34.65	20.20				

These results, having established rather conclusively that the quantity of sand crushed under the conditions of the test bore no relation to the relative priming efficiencies, even of such similar explosive mixtures as mercury fulminate and oxygen carriers, an explanation of the interpretation to be placed upon the "sand test" was sought.

In Table 12, following, the explosives are listed in the order of their calculated total energy of explosion. The second column shows the calculated heat developed by the explosion of 1 gram of each explosive, the value being based on Martin's value of 349 calories for 1 gram of mercury fulminate. The heats of reaction of the salts used in the calculation were taken from Biedermann's Chemiker Kalender for 1916. Assuming that the specific heats of the products of explosion are approximately equal, the total energy of explosion may be expressed by Berthelot's "characteristic product," QV_o, in which V_o is the calculated volume of gaseous products of explosion at 0° C. and 760 mm. pressure. Columns 5 and 6 give the results of "sand tests." It will be noted that the values for the "sand tests" stand in nearly the same order as the product QV_o, indicating that the sand test is really a test for strength. This conclusion suggests that a sand test modified for the brisant blasting explosives and on a much larger scale might give more satisfactory results than either the Trauzl lead block or the ballistic mortar.

TABLE 12.—*Energy of priming compositions.*

Primer mixed with mercury fulminate.	Heat of explosion of 1 gram of explosive, calories.	Volume of gas at 0° C. and 760 mm., c. c.	Characteristic product.	Results of sand tests.	
				Shell not pressed.	Shell pressed under pressure of 200 atmospheres per square inch.
	Q	V_o.	QV_o.		
PbO$_2$, 20 per cent...............	345	252	86,000	32.80	32.20
PbO$_2$, 10 per cent...............	342	283	97,500	34.40	33.45
Mercury fulminate...............	349	315	110,000	36.10	35.30
KMnO$_4$, 20 per cent............	473	238	112,500	34.40	34.00
KMnO$_4$, 10 per cent............	411	276	113,500	34.90	34.60
Pb(NO$_3$)$_2$, 10 per cent...........	406	290	117,500	36.50	36.35
Pb(NO$_3$)$_2$, 20 per cent...........	463	266	123,000	37.40	35.60
NaNO$_3$, 10 per cent.............	488	283	138,000	38.05	38.80
KClO$_3$, 10 per cent.............	490	283	138,500	38.75	36.95
Ba(ClO$_3$)$_2$.H$_2$O, 10 per cent......	487	290	141,000	39.10	36.80
KClO$_4$, 10 per cent.............	506	283	143,000	39.30	38.40
NaNO$_3$, 20 per cent.............	610	252	153,500	38.90	37.55
KClO$_3$, 20 per cent.............	632	252	159,000	39.90	39.25
KClO$_4$, 20 per cent.............	652	252	163,500	41.00	41.20
Ba(ClO$_3$)$_2$.H$_2$O, 20 per cent......	626	266	166,500	39.35	38.70

DISCUSSION OF THEORY.

Until more is known concerning the process called "detonation" in high explosives, and more definite scientific concepts than percussion, blow, impact, shock, etc., are formulated for describing the initial impulse that causes high explosives to detonate, it will be difficult to determine the peculiar properties upon which the efficiency of initial priming explosives depend.

Accepting tentatively the views of Wöhler (31), Storm and Cope (27) consider that the initial velocity, or the initial rate of detonation, is one of the factors of greatest importance in determining the efficiency of priming substances. As Marshall (18) has pointed out, Wöhler really means acceleration and not initial velocity (Anfangsgeschwindigkeit). When a priming explosive is ignited by flame it first burns, but in an almost infinitesimally small space of time detonation sets in. In other words, the more efficient primers accelerate their own explosive transformation enormously. The less efficient primers can be accelerated by priming them with more efficient primers, for example, lead hydronitride on mercury fulminate. It may not be going too far even to say that theoretically explosives of the second order require primers simply because they lack the capacity to accelerate burning through the intermediate stages to the detonation stage.

The effectiveness of a detonator in practice probably depends upon explosive energy that may be separated into an intensity factor and a capacity factor.[a] The intensity factor may be called "quickness," and depends upon acceleration of the explosive decomposition. The capacity factor is "strength" and is measured by the total energy of the primer $\dfrac{QV_o}{c}$ in which Q is the heat of the reaction, V_o the volume of gas liberated, and c the specific heat of the products of reaction.[b] The "sand test" may be regarded as a measure of "strength." The lead-plate test and the shattering effects on glass tubes are tests of "quickness." Of the two factors the quickness is considered the more important (23), but as this may be regarded as constant for a given primer, the strength test (sand test) is a measure of the efficiency of detonators containing the same composition.

The priming properties of the particular sample of silver acetylide used in the investigation affords a striking example of the importance of both the quickness and the strength of primers. This compound was much inferior to mercury fulminate as a primer for TNT and tetryl, yet it was quick enough to accelerate mercury fulminate and to increase considerably the efficiency of the latter (see Table 9). It gave a very inferior sand test (see Table 11), which probably accounts for its inefficiency as a primer for the nitro compounds. On the other hand, lead hydronitride, which increases the efficiency of fulminate as a primer for TNT, could not accelerate the silver acetylide sufficiently to overcome its lack of strength.

---

[a] "Intensity" and "capacity" are not used here in exactly the most accurate way. See Noyes, A. A., General principles of physical science, 1902, p. 79–82.

[b] $\dfrac{QV_o}{c}$ is Berthelot's "characteristic product." See page 21.

VALUE OF SAND TEST.

The fact that the efficiency of initial priming explosives for detonating such typical high explosives as TNT and tetryl is not paralleled by their ability to crush quartz sand in the sand bomb does not impair the usefulness of the latter for testing commercial detonators. The quantitative character of the results gives the sand test a decided advantage over other tests, and if the detonators loaded with the same composition are considered, it is an accurate index of their relative efficiencies. It is planned to correlate the results of the sand test with tests of cartridges of blasting, mining, and other high explosives, but until this can be done detonator compositions of radically different character should also be subjected to other tests, especially those depending on the quickness of the priming.

As the priming efficiency will probably be found to depend upon both factors of "quickness" and of "strength," the value of the sand test for determining at least the "strength" may be considered as established.

WORKS CONTROL.

The value of the sand test for such investigations as are described in this report is unquestionable. It indicates accurately whether the nitro compound comprising the base charge of the detonator has completely detonated, partly detonated, or failed altogether.

In the manufacture of mercury fulminate, or other initial priming explosives, some method of testing as an aid in obtaining uniformity of product is desirable. Chemical analysis is one such method. Determination of the minimum charge required to detonate a standard nitro compound, such as trinitrotoluene, for example, according to the method described in Bureau of Mines Technical Paper 145,[a] is suggested as another useful method. By this method a small charge of fulminate, its chlorate mixtures, or lead hydronitride is superimposed upon the nitro compound in the detonator shell. The detonator is usually of the "reinforced" type, in which an inner perforated capsule is pressed down over the charge. The sensitive fulminate composition, fired by fuse or electrically, detonates the secondary explosive, thereby transmitting the impulse that causes the blasting explosive to detonate. The tests are usually carried out by weighing into the copper shell a fixed quantity of explosive and varying the weight of priming until the minimum quantity required for certain detonation is ascertained. Variations in the manufacture, mixing, pressing, and keeping of compositions for detonators could easily be detected by this means and the entire

---

[a] Taylor, G. B., and Cope, W. C., Sensitiveness to detonation of trinitrotoluene and tetranitromethylanilin : Tech. Paper 145, Bureau of Mines, 1916, p. 5.

manufacturing operations subjected to rigid control by it. The tests do not require the service of especially trained chemists, as the only expert operation needed is careful weighing of the detonator charges.

ACKNOWLEDGMENTS.

The authors desire to acknowledge the valuable assistance rendered by the following Bureau of Mines employees: J. Barab, junior explosives engineer, who prepared the organic priming compounds, and J. E. Crawshaw, junior explosives engineer (resigned), and V. P. Hawse, junior chemist, in preparing the acetylides. C. Matthews. junior explosives engineer (resigned), and G. F. Hutchison, junior chemist, assisted in making some of the sand tests.

SELECTED BIBLIOGRAPHY.

HISTORICAL REPORTS.

1. Escales, R., Die explosivstoffe. Leipsic, 1908. Ht. 3, pp. 1–46 and 161–163.
2. Howard, E., On a new fulminating mercury: Philos. Trans., vol. 90, 1800, p. 204.
3. McDonald, G. W., The discovery of fulminate: Arms and Explosives, vol. 19, 1911, p. 24.
4.——— Historical papers on modern explosives. New York, 1912. 192 pp.
5. de Mosenthal, H., Life work of Alfred Nobel: Jour. Soc. Chem. Ind., vol. 18, 1899, pp. 443–451.

PROPERTIES AND ACTION OF PRIMING EXPLOSIVES.

6. Abel, F., Contribution to the history of explosive agents: Philos. Trans., vol. 159, 1869, pp. 489–516; Compt. rend., t. 69, 1869, pp. 105–121. Explains theory of wave synchronism.
7. Anonymous. (Detonation of wet guncotton charges): Sprengstoffe, Waffen, und Munition, Bd. 3, 1908, p. 277.
8. ——— Detonation of explosives in high-explosive shells: Engineering, vol. 72, 1901, pp. 156, 195.
9. ——— (Explosibility of mercury fulminate): Sprengstoffe, Waffen, und Munition, Bd. 4, 1909, p. 270. Describes conditions under which fulminate explodes by heat, blow, etc.
10. Berthelot, M. P. E., Explosives and their power (English translation by Hake and Macnab): London, 1892, pp. 563.
11. ———, Sur l'explosion du chlorate de potasse (Explosion of potassium chlorate): Mem. Poud. et Salpetres, t. 10, 1900, pp. 280–283; Jour. Soc. Chem. Ind., vol. 20, 1901, p. 388. Treats of detonation of pure potassium chlorate brought about by sudden application of heat.
12. Brunswig, H., Explosives (translation by Munroe and Kibler). 1912, chs. 1, 2, and 8, pp. 1–106, 219–234.
13. ———, Neue Initialzündung für Sprengstoffe: 8th Int. Cong. App. Chem., Original communications, vol. 4, pp. 19–22. Describes " loop " detonating fuse.
14. Escales, Richard, Die explosivstoffe. Leipsic, 1908. Ht. 3, pp. 161–163. Initial primers for nitroglycerin explosives.
15. Hess, P., Neurungen im Spreng und Zündmittelwesen: Ztschr. angew. Chem., Jahrg. 17, 1904, pp. 545–554. A lecture on explosives and priming agents.
16. Kibler, A. L., Hydronitrides: 8th Int. Cong. App. Chem. (Appendix), Original communications, vol. 25, pp. 235–238.
17. Le Chatelier, H., Explosifs: Rev. métall., t. 12, January, 1915, p. 69. General article on priming compositions.

18. MARSHALL, A., Explosives. London, 1915, ch. 29, Ignition, pp. 414–434. Short account of recent proposals and theories of detonation. Describes manufacture of firearms, primers, and blasting caps.

19. MARTIN, F., Über Azide and Fulminate. Darmstadt, 1913.

20. MEYER, V., Beobachtungen Vermischten Inhalts; Beobachtungen am Dynamit: Liebig's Annalen, Bd. 264, 1891, p. 127. A criticism of Abel's hypothesis on action of primers.

21. NEITZEL, —., Die Sprengstofftechnik der Initialzündungen: Ztschr. ges. Schiess-Sprengstoffw., Bd. 8, 1913, pp. 145, 167, 190, 209, 231. Excellent article on recent theories and methods of detonation and priming.

22. PETER, A. M., Stability of silver fulminate: Jour. Am. Chem. Soc., vol. 38, 1916, p. 486.

23. STETTBACHER, A., Altes und Neues uber Initialzündstoffe: Ztschr. ges. Schiess-Sprengstoffw., Bd. 9, 1914, pp. 341, 355, 381, 391. Excellent discussion of recent theories on the priming action of initiating explosives.

24. ———, Vergleichende Explosionswirkungen (Comparative action of explosives) : Ztschr. ges. Schiess-Sprengstoffw., Bd. 10, 1915, pp. 193, 214. Describes mechanical effect of initial explosives on an iron base.

25. ———, Verfahren und vorschlage zum Detonieren von Sprengladungen: Ztschr. ges. Schiess-Sprengstoffw., Bd. 10, 1915, pp. 16–19. Discusses advantages of core of detonating fuse through the charge and detonation by the " loop " detonating fuse.

26. ———, Neuere initial Explosivstoffe (Recent priming explosives) : Ztschr. ges. Schiess-Sprengstoffw., Bd. 11, 1916, pp. 1–3.

27. STORM, C. G., and COPE, W. C., The sand test for determining the strength of detonators: Tech. Paper 125, Bureau of Mines, 1916, 68 pp.

28. STUART, A. J., Lead hydronitride: Jour. U. S. Artillery, vol. 41, 1914, pp. 212–223. Translation of a pamphlet by the Rheinisch-Westfälische Sprengstoffe Aktien-Gesellschaft, of Cologne, Germany, comparing the properties of mercury fulminate and lead azide (hydronitride).

29. THRELFALL, R., On the theory of explosives: Philos. Mag., vol. 21, 1886, pp. 165–179.

30. WILL, W., Der Fortschritt in der Auslosung der Explosiblen Systeme und sein Einfluss auf die Sprengstoffindustrie: Ztschr. ges. Schiess-Sprengstoffw., Bd. 9, 1914, pp. 52–53. Lecture including the general subject of detonators.

31. WÖHLER, L., Über Initialzündung: Ztschr. ges. Schiess-Sprengstoffw., Bd. 6, 1911, p. 253; Ztschr. angew. Chem., Jahrg. 24, 1911, pp. 1111, 2089–2098. Lecture before the Chemical Society at Stettin, setting forth author's theory of initial primers.

32. WÖHLER, L., and MARTIN, F., Neue Salze der Knallsäure und Stickstoffsäure und die Explosives Eigenschaften von Fulminaten und Aziden: Ztschr. angew. Chem., Jahrg. 27, 1914, pp. 335–6. Methods of preparation, properties, and explosive action of K, Ca, Sr, Ba, Mn, Zn, Te, and Cd fulminates and Mn, Zn, and Co nitrides.

33. WÖHLER, L., and MATTER, O., Beitrag zur Wirkung der Initialzündung von Sprengstoffen: Ztschr. ges. Schiess-Sprengstoffw., Bd. 2, 1907, pp. 181, 203, 244, and 265. Examination of seven explosives for priming effect and comparison of explosives properties with mercury fulminate, showing that initial primers more efficient than the latter exist, which the authors consider overturns Abel's hypothesis.

34. WOLF, P., Erzstzstoffe für Knallquicksilber in Zündsäten (Substitues for mercury fulminate in priming compositions) : Ztschr. ges. Schiess-Sprengstoffw., Bd. 11, 1916, p. 4.

PREPARATION, CONSTITUTION, ETC.

35. BROWNE, A. W., A new synthesis of hydronitric acid: Jour. Am. Chem. Soc., vol. 27, 1905, pp. 551–555.

36. BROWNE, A. W., and HOULEHAN, A. E., Behavior of hydronitrogens and their derivatives in liquid ammonia; (2) Ammonolysis of certain hydrazine salts: Jour. Am. Chem. Soc., vol. 33, 1911, pp. 1734–1742.

37. ——, Behavior of hydronitrogens and their derivatives in liquid ammonia; (3) Action of ammonium trinitride upon certain metals: Jour. Am. Chem. Soc., vol. 33, 1911, pp. 1742–1752.

38. ——, Behavior of hydronitrogens and their derivatives in liquid ammonia; (4) Pressure-concentration isotherms in the system ammonia; ammoniumtrinitride: Jour. Am. Chem. Soc., vol. 35, 1913, pp. 649–658.

39. BROWNE, A. W., and HOLMES, M. E., Behavior of hydronitrogens and their derivatives in liquid ammonia; (5) Electrolysis of a solution of ammonium trinitride in liquid ammonia: Jour. Am. Chem. Soc., vol. 35, 1913, pp. 672–681.

40. BROWNE, A. W., and LUNDELL, G. A., Anhydrous hydronitric acid; (1) Electrolysis of a solution of potassium trinitride in hydronitric acid: Jour. Am. Chem. Soc., vol. 31, 1909, pp. 435–448.

41. BROWNE, A. W., and WELSH, T. W., Behavior of hydronitrogens and their derivatives in liquid ammonia; (1) Ammonolysis of hydrazine sulphate: Jour. Am. Chem. Soc., vol. 33, 1911, pp. 1728–1734.

42. CURTIUS, THEODORE, Neues vom Stickstoffwasserstoff: Ber. deut. chem. Gesell., Jahrg. 24, 1891, pp. 3341–3349. Preparation and properties of hydronitrides.

43. CURTIUS, THEODORE, and RISSOM, J., Neue Untersuchungen über Stickstoffwasserstoff N_3H: Jour. prakt. Chem., Bd. 58, 1898, pp. 261–309.

44. DARAPSKY, A., Die Salze der Stickstoffwasserstoffsäüre als Explosivstoffe: Ztschr. ges. Schiess-Sprengstoffw., Bd. 2, 1907, pp. 41 and 64.

45. DENNIS, L. M., Hydronitric acid and hydronitrides: Jour. Am. Chem. Soc., vol. 18, 1896, p. 947.

46. DENNIS, L. M., and BENEDICT, C. H., Upon the salts of hydronitric acid: Jour. Am. Chem. Soc., vol. 20, 1898, pp. 225–232.

47. DENNIS, L. M., and BROWNE, A. W., Hydronitric acid and the inorganic trinitrides: Jour. Am. Chem. Soc., vol. 26, 1904, pp. 577–612.

48. DENNIS, L. M., and DOAN, M., Some new compounds of thallium: Jour. Am. Chem. Soc., vol. 18, 1896, pp. 970–977.

49. DENNIS, L. M., and ISHAM, H., Hydronitric acid: Jour. Am. Chem. Soc., vol. 29, 1907, pp. 18–31, 216–223.

50. GIRSEWALD, C. V., Beiträge zur Kenntnis des Wasserstoffperoxyds; Üeber die Einwirkung des Wasserstoffperoxyds auf Hexamethylentetramin (The action of hydrogen peroxide on hexamethylentetramine): Ber. deut. chem. Gesell., Jahrg. 45, 1912, p. 2571.

51. HERZ, E., Perchloric diazo compounds: German patent No. 258679; Chem. Abs., vol. 7, 1913, p. 2687; U. S. patent 1064411, February 25, 1913.

52. HOFMAN, K. A., Explosive Quecksilbersalze: Ber. Deut. chem. Gesell., Jahrg. 38, 1905, pp. 1999–2005. Mercuraldehydes of chloric and perchloric acids.

53. KIBLER, A. L., Mercury fulminate from propyl alcohol: 8th Int. Cong. App. Chem. (Appendix), Original communications, vol. 25, pp. 239–243.

54. MUNROE, C. E., Production of mercury fulminate: Jour. Ind. and Eng. Chem., vol. 4, 1912, pp. 152–153. Note on substituted fulminic acids from higher alcohols.

55. Nef, J. U., Ueber die Constitution der Salze der Nitroparaffine: Liebig's Annalen, Bd. 279–280, 1886, pp. 263–291.

56. ———, Ueber zweiwerthige Kohlenstoffatom; Liebig's Annalen, Bd. 280, 1886, pp. 291–342.

57. Philip, R., Knallquecksilberstudien: Ztscher. ges. Schiess-Sprengstoffw., Bd. 7, 1912, pp. 109, 156, 180, 198, 221. Analysis establishing difference between white and gray samples.

58. Schenck, R., Über den Schwefelstickstoff: Liebig's Annalen, Bd., 290, 1896, pp. 171–185.

59. Solonina, A., Das Knallquecksilber: Ztschr. ges. Schiess-Sprengstoffw., Bd. 5, 1910, pp. 41–46.

60. Szuhay, J., Beiträge zur Kenntnis des Jodstickstoff (Nitrogen iodide): Ber. Deut. chem. Gesell., Jahrg. 26, 1893, p. 1933.

61. Thiele, J., Uber die Konstitution der aliphatischen Diazoverbindungen und der Stickstoffwasserstoffsäure. Ber. Deut. chem. Gesell., Jahrg. 44, 1911, pp. 2522–2525.

62. Wieland, H., Die Knallsäure (Enke, Stuttgart).

63. Wieland, H., and Bauman, A., Zur Kenntnis der polymeren Knallsäuren (Polymeric fulminic acids): Liebig's Annalen, Bd. 392, 1912, pp. 196–213.

64. Wöhler, L., Die Molekulargrosse der Knallsäure (Molecular weight of fulminic acid): Ber. Deut. chem. Gesell., Jahrg. 38, 1905, pp. 1351–1359.

65. Wöhler, L., and Krupko, W., Über die Lichtempfindlichkeit der Azide des Silbers, Quecksilberoxyduls, Bleis, und Kupferoxyduls, sowie über basische Blei und Cupriazid (sensitiveness to light of azides, etc.): Ber. Deut. chem. Gesell., Jahrg. 46, 1913, pp. 2045–2057.

66. Wöhler, L., and Theodorovitz, K., Beitrag zur Aufklärung des Knallquecksilberprocesses: Ber. Deut. chem. Gesell., Jahrg. 38, 1905, pp. 1345–1351.

67. Wolf, P., Reinigen des Knallquecksilbers (Purification of mercury fulminate): Ztscher. ges. Schiess-Sprengstoffw., Bd. 7, 1912, p. 272.

MANUFACTURE AND TESTS OF DETONATORS.

68. Guttman, O., Manufacture of explosives. London, 1895. 2 vols.

69. Hagen, O., Ein Beitrag zur Fabrikation des Knallquecksilbers (Manufacture of mercury fulminate and detonators): Ztscher. ges. Schiess-Sprengstoffw., Bd. 6, 1911, pp. 4, 28, 201, 224, 243, 265, 283, 308. Discusses modern practice abroad.

70. Hall, C. and Howell, S. P., Investigations of detonators and electric detonators: Bull 59, Bureau of Mines, 1913, 73 pp. Describes direct and indirect engineering tests.

71. Knoll, R., Das Knallquecksilber und ähnliche Sprengstoffe: A. Hartleben's Chemische techische Bibliothek, Vienna and Leipsic, 1908. A treatise on the manufacture of mercury fulminate.

72. Marshall, A., Explosives. London, 1915, 624 pp. Describes manufacture.

73. Oliver, R. L., Detonating caps for blasting: Eng. and Min. Jour., vol. 82, 1906, pp. 682–686. Practical discussion of reasons for failure in the field.

74. Reifsneider, L. B., Comparative strength of detonators: Eng. and Min. Jour., vol. 99, 1915, p. 818. Discusses effect of strength of detonator on explosion of dynamite by influence.

75. Taffanel, J., and Dautriche, H., Sur le mode d'amoreage des explosifs (Methods of ignition of explosives): Compt. rend., t. 153, 1911, pp. 823–825. Discusses effect of position of detonator.

RECENT PATENTS.

76. ALVISI, U., Mercury fulminate mixed with ammonium perchlorate: German patent 124103, June, 10, 1898; British patent 25838, December 17, 1898; United States patent 707493, August 19. 1902.

77. BICKFORD-SMITH, G. P., Composition for blasting cap: British patent 20989. October 5, 1898.

78. BRUNSWIG, H., Neue Initialzundung für Sprengstoffe (Loop detonating fuse): 8th Inter. Cong. Appl. Chem., Original communication, vol. 4, pp. 19–22; German patent 245087. September 23, 1910; British patent 21344, September 27, 1911; United States patent 1042643 October 29, 1912.

79. BURKARD, E., Detonating fuse. Trinitroethyaniline and trinitrochlorbenzene: United States patent 1049665, January 7, 1913; United States patent 1049666, January 7, 1913.

80. CALVET, R., Persulphocyanic acid, heavy metal salts of: German patent 263231, June, 12. 1912; British patent 9597, June 11, 1913; French patent 459014, May 24, 1913; Swiss patent 62590, May 27, 1913.

81. CLAESSEN, C., Tetranitromethylaniline and fulminate of mercury: German patent 166804, March 18, 1905; British patent 13340, 1905; French patent 355695, June 28, 1905; United States patent 827768, August 7, 1906.

82. ———, Tetranitroethylaniline and fulminate of mercury: German patent 168490, April 7, 1905; British patent 13340, 1905; French patent 355695, June 28, 1905.

83. ———, Nitropentaerythritol: German patent 265025, December 8, 1912; British patent 29901, December 28, 1912; Austrian patent 64976, December 15, 1913; French patent 451925, December 14, 1912; Swiss patent 61926, December 7, 1912.

84. ———, Loop detonating fuse: German patent 245087, September 23, 1910; British patent 21344, September 27, 1911.

85. ———, Lead azide used with fulminate: French patent 459979, May 29, 1913; first addition July 7, 1913; British patent 13086, June 5, 1913; Swedish patent 40379, March 15, 1916.

86. ———, Priming lead azide with black powder, guncotton, etc.: British patent 16456, July 17, 1913.

87. ———, Hexanitroethane and fulminate of mercury: French patent 463714, October 17, 1913.

88. FAIRWEATHER, W., Highly compressed tetranitromethyl and ethyl anilines primed with lead azide: British patent 2407, January 30, 1912.

89. FÜHRER, J., Copper ammonium nitrate, potassium nitrate, sulphur, and aluminum, in hermetically sealed container for high explosives: British patent 2755, October 16, 1901.

90. GEHRE, F., Di and trinitromesitylene, pseudocumene, and xylene with primer of fulminate: British patent 19402, September 26, 1905.

91. VON GIRSEWALD, Conway, Verwendung von Hexamethylentriperoxidiamin zur Hertsellung von Initialzündern: German patent 274522, September 14, 1912.

92. HALE, C. M., and BELL, M., Lead picrate: British patent 13348, June 12, 1902.

93. HARLÉ, J., Tetranitropentaerythritol alone or mixed with nitro compounds for detonating fuse: British patent 15355, June 26, 1914.

94. HERZ, E., Nitrodiazobenzene perchlorate (and homologs of benzene): German patent 258679, April 27, 1911; British patent 27198, November 26, 1912; French patent 450897, November 21, 1912; United States patent 1054411, February 25, 1913.

95. ———, Lead salt of trinitroresorcin: British patent 17961, October 29, 1915.

96. HESS, P., Dephlegmatizing mercury fulminate with glycerin, rosin, etc.: British patent 3238, February 8, 1902.

97. HYRONIMUS, F., Lead azide: German patent 224669, January 26, 1908; Austrian patent 41890, November 15, 1909; United States patent 908674, January 5, 1909; French patent 384792, February 14, 1907.

98. JAQUES, A., Benzoyl peroxide: British patent 23450, October 14, 1912.

99. LHEURE, LOUIS, Detonating fuse of trinitrotoluene: United States patent 869219, October 22, 1907.

100. MAIN, W. L., Fulminate mixed with bromates: United States patent 1147958, July 27, 1915.

101. NEWTON, A. V. (from A. Nobel, Paris), Granulated mixture of lead picrate, potassium picrate, and potassium chlorate: British patent 16919, December 8, 1887.

102. RHEINISCH-WESTFÄLISCHE SPRENGSTOFF-ACT.-GES. IN CÖLN, Initialzünder (Azides of heavy metals with fulminates, or substances like nitrogen sulphide and diazobenzolnitrate); German patent 238942, June 30, 1910; Austrian patent 60852, April 15, 1913; Swiss patent 56277, March 15, 1911.

103. RUNGE, W., Mixtures of fulminates, azides, etc., with aromatic nitrohydocarbons: United States patent 1168746, January 18, 1916.

104. VON SCHROETTER, O., Hexanitrodiphenylamine: United States patent 908149, December 29, 1908.

105. ———, Nitro compounds used in detonators for moist nitrocellulose explosives: French patent 376339, April 3, 1907; British patent 8157, April 11, 1907.

106. SCHWARZ, A. W., Nitrogen iodide, chloride, or bromide: British patent 7098, April 10, 1894.

107. SOCIÉTÉ ANONYME DYNAMITE NOBEL, Silver acetylide: French patent 321285, May 21, 1902.

108. SPRENGSTOFFE AKTIEN-GESELLSCHAFT CARBONIT, Hexanitrodiphenyl: British patent 18333, August 7, 1914.

109. SPRENGSTOFF-FABRIKEN HOPPECKE, ACT.-GES. IN HAMBURG (Blasting cap with nitrocellulose bottom): German patent 182985, October 19, 1905.

110. VENIER, W., Mercury acetylide: French patent 364461, March 21, 1906; British patent 6705, March 20, 1906.

111. WESTFÄLISCHE ANHALT SPRENGSTOFFE AKTIEN-GESELLSCHAFT, Di or tri nitrocresol alone or mixed with potassium chlorate: French patent 320199, April 4, 1902.

112. WHITE, W. C., Mercury fulminate, 87.5 per cent, with silver fulminate, 12.5 per cent: British patent 13983, June 19, 1906; British patent 20439, September 13, 1907.

113. WILL, K. W., Tetranitro methyl and ethyl aniline with mercury fulminate primer: United States patent 827768, August 7, 1906.

114. WÖHLER, L., Nitro derivatives primed with fulminate: British patent 21065, November 21, 1900.

115. ———, Silver, lead, and mercury azides, etc.: German patent 196824, March 2, 1907; British patent 4468, February 27, 1908; Austrian patent 37029, December 1, 1908; French patent 387640, February 28, 1908; French patent 14485 (addition), June 27, 1911; United States patent 904289, February 17, 1908; Swiss patent 45491, June 9, 1908; United States patent 1128394, February 16, 1915.

www.ingramcontent.com/pod-product-compliance
Lightning Source LLC
Chambersburg PA
CBHW021433180326
41458CB00001B/246